A REVIEW OF NUCLEAR ENERGY
IN THE
UNITED STATES

A REVIEW OF NUCLEAR ENERGY IN THE UNITED STATES

Hidden Power

TODD H. OTIS

PRAEGER

PRAEGER SPECIAL STUDIES • PRAEGER SCIENTIFIC

Library of Congress Cataloging in Publication Data

Otis, Todd H.
   A review of nuclear energy in the United States.

   Bibliography: p.
   Includes index.
   1. Atomic power industry—United States.. I. Title.
HD9698.U52087   338.4′762148′0973   81-11859
ISBN: 0-03-060001-4   AACR2

Published in 1981 by Praeger Publishers
CBS Educational and Professional Publishing
A Division of CBS, Inc.
521 Fifth Avenue, New York, New York 10175 U.S.A.

© 1981 by Praeger Publishers

All rights reserved

23456789 052 98765432

Printed in the United States of America

To Jane, my wife, who kept the faith.

ACKNOWLEDGMENTS

I would like to thank the people who helped set off the chain of events that led me into this fascinating inquiry and those who helped me understand and explain some of the technical issues related to nuclear power: Dean Abrahamson; an anonymous employee of the U.S. Department of Energy; Tom Donovan; John Gostovich; Byron Harris; Steve Leuthold; David McDermott; Saunders Miller; Emily Moore; Susan Nelson; Louise Nichols; Harold Nicholson; Ken Peterson; Winthrop Rockwell; Barbara White; and Representatives Irv Anderson, Phyllis Kahn, and Ken Nelson, all from the Minnesota House of Representatives.

I am grateful as well to Cheri Abernathy for her excellent typing work and to Jane Ferguson for her fine editing.

CONTENTS

	Page
ACKNOWLEDGMENTS	vi
LIST OF TABLES AND FIGURES	x
INTRODUCTION	xi

Chapter

1 A TRIP TO MONTICELLO AND BEYOND — 1

    On the Low Volume of Commercial High-Level Waste — 3
    On Whether the Disposal Problem Is Mainly Political, Not Technical — 7
    On the Cost of Disposing of Highly Radioactive Waste — 15
    Conclusions and Comments — 17
    Notes — 18
    References — 20

2 THE SURPRISING ECONOMICS OF NUCLEAR POWER — 23

    Rising Cost of Nuclear Power — 26
    A Case Study: Proposed Tyrone Nuclear Power Plant — 32
    Hidden Costs of Nuclear Power — 36
    The Batelle Study — 37
    Additional Subsidies and Hidden Costs — 43
    The Costs of an Accident — 44
    Conclusions and Comments — 47
    Notes — 49
    References — 51

3 LOW-LEVEL RADIATION: HIDDEN DANGER — 55

    Low-Level Radiation: How Safe? — 56
    Radiation Dangers in the Nuclear-Fuel Cycle — 61
    Conclusions and Comments — 68

Chapter		Page
	Notes	69
	References	71
4	REACTOR ACCIDENTS: CLEAR AND PRESENT DANGER	73
	An Overview of Nuclear Reactors	74
	Reactor Accidents: They Happen	77
	Preventing and Managing Accidents	84
	The NRC as Regulator: How Tough—How Fair?	87
	Catastrophes: They Are Not Supposed to Happen	92
	The Rasmussen Report and the Lewis Report	93
	Conclusions and Comments	95
	Notes	97
	References	99
5	HOPEFUL ALTERNATIVES: THE FUTURE IS NOW	101
	The Technology of Conservation and Energy Efficiency	103
	The Wisdom of Conservation for Utilities	112
	Cogeneration	118
	Wind Power	120
	Hydroelectric Power	122
	Biomass	124
	Geothermal Energy	125
	Photovoltaic Electricity	126
	Centralized Solar Power	128
	Centralized Wind Power	129
	Power from the Sea	129
	A Strategy to Phase Out Nuclear Power	130
	Conclusions and Comments	136
	Notes	137
	References	141
6	HIDDEN POWER	145
	Selling the Idea of Nuclear Power	147
	Buying the Idea of Nuclear Power	152
	Ending Hidden Power	154
	Notes	156
	References	157

Appendix		Page
A	A LOOK AT ATOMS AND RADIOACTIVE WASTE	159
B	UTILITY REGULATION AND RATE MAKING	161
C	IONIZING RADIATION AND ITS EFFECTS ON CELLS	165
INDEX		171
ABOUT THE AUTHOR		176

LIST OF TABLES AND FIGURES

Table		Page
2.1	Federal Funding of Commercial Nuclear Power, 1950-77	38
2.2	Wall Street Nuclear Fallout, Performance Comparisons	47
4.1	List of Abnormal Occurrences at Nuclear Power Plants	78
5.1	Potential for Meeting 1987 Projected Demand without Nuclear Power	134

Figure		
1.1	Estimated Civilian and Military High-Level Waste Inventories Measured in Terms of Their Strontium 90 Content as a Function of Time	5
2.1	Regulatory Growth of the Nuclear Industry	27
4.1	Boiling Water Reactor (BWR)	75
4.2	Pressurized Water Reactor (PWR)	76
5.1	Nuclear-Power-Plant Locations and Interregional Power-Transfer Capabilities	132
6.1	Middle South Utilities Advertisement	148

INTRODUCTION

Electricity is almost like magic. With the flick of a switch a dark room is illuminated, a hot room is cooled, a lonely room is visited by Johnny Carson. Without the need to understand how it happens, people have been able to use electricity to do almost anything from brushing their teeth to vibrating a bed. It provides the most immediate kind of power, a handy power that nearly every one of us enjoys. How we continue to generate and use that electricity is what this book is about; its focus is nuclear power.

I was a freshman legislator in the Minnesota House of Representatives, a Democrat elected about five months before the accident at Three Mile Island. I had been advised by a respected legislator who had just retired from the House to avoid carrying any controversial legislation as a freshman. It was good advice, since understanding the complexities of procedure and understanding proposed bills well enough to vote intelligently offer ample challenge for any legislator's first two years. However, I did agree to carry a bill dealing with nuclear waste, which was presented to me in early February 1979. The representative in the House usually associated with such an issue had thought it a good idea for a new person to oversee this kind of legislation, since she was viewed as an outspoken liberal and the 1978 elections in Minnesota had brought a major influx of conservative Republicans. I was viewed as more neutral on issues such as nuclear power and less likely to present a red flag to the legislators who would be voting on such legislation.

The bill I agreed to sponsor simply provided that no new nuclear power plants be allowed to be built Minnesota until a safe and economically feasible means of disposing of the waste could be found. Although it would later be labeled an "ironclad moratorium," it merely established a responsible precondition for the future use of nuclear power in the state. I knew very little about nuclear power at the time, but on the surface this bill (House File 378, as it was to become officially known) seemed moderate and reasonable—not extreme or dangerous. Indeed, when I agreed to be chief author of HF 378, I was certainly not antinuclear. I did not know much about the technical complexities or economics of this energy form. I did feel that it was irresponsible to allow tons of highly radioactive material to accumulate without knowing what to do with it on a long-term basis, but I felt that while nuclear power was somewhat dangerous, it was cheap and probably necessary.

One of the main hurdles I had to overcome in carrying HF 378 was a sense of intimidation by the technical nature of nuclear power. I knew little of the workings of atoms and how nuclear fission comes about, creating as it does the extraordinary heat and energy in nuclear reactors. Moreover, I was unversed in the economics of nuclear power, as well as unfamiliar with radiation, reactor safety, and what realistic alternatives to this source of energy existed. I had a great deal to learn.

Getting good information on nuclear power and nuclear waste became my central concern as HF 378 worked its way through the legislative process. The main opponent of the bill was Northern States Power Company (NSP), an investor-owned utility that generates 43 percent of its electricity in Minnesota by means of nuclear power. The Sierra Club and the Minnesota Public Interest Research Group (MPIRG), a college-student-funded organization that researches and advocates changes in laws that affect the environment, combined to present the case against further development of nuclear power until the waste-disposal issue could be resolved. It had been MPIRG that initiated my bill in the first place.

NSP presented nuclear power information to the Energy Committee, to which the bill was referred, in three ways: it gave the committee a tour of its Monticello nuclear power plant, it distributed a fact sheet on HF 378, and it testified against the bill at committee hearings in 1979 and 1980.

Although I sought information from NSP on several technical questions related to nuclear power and met with patient and gracious cooperation, I felt compelled to look beyond that company for information. Indeed, it was the follow-up research I did on NSP's assertions that led me to write this book.

Among the statements made by NSP in the fact sheet they circulated in the late winter of 1979 were these assertions:

> Nuclear plant electricity is cheap.
>
> Ultimate disposal of nuclear wastes is feasible.
>
> Ultimate nuclear waste disposal costs can be predicted.
>
> California, which passed a similar bill, has begun seeking federal permission to allow burning of oil to meet their electricity needs.

> Conservation and vigorous development of conventional and alternative energy sources will not meet the energy needs into the next century without nuclear power.

In the testimony against the bill in April 1979, NSP's nuclear expert stated, among other things, the following:

> Effective development of coal as an energy source has been slow in coming, and other alternate energy sources are probably decades away, in terms of electrical generation.
>
> Nuclear plant electricity is less expensive.
>
> This [HF 378] will leave Minnesota hostage to power supplied by coal and oil—subject to strikes, price hikes, transportation problems, and so on.

To be sure, there were countervailing statements and arguments brought to bear by supporters of HF 378. In my summary to the Utilities Subcommittee of the House Energy and Utilities Committee in 1979, I pointed out that the same NSP expert who had told the committee that nuclear power was less expensive than coal when he testified on the bill had characterized coal and nuclear power costs as a "wash" when the committee had toured the Monticello nuclear plant; that estimating the cost of nuclear-waste disposal without knowing what technology was going to be used or where it would be located was on the surface absurdly speculative; and that California had been burning oil for years to generate its electricity.

The bill failed to pass out of subcommittee on a six-to-six, party-line tie vote in April 1979. A special election held that summer tipped the balance of power in the House of Representatives to the Democrats, and the bill was heard again and passed out of the full Energy Committee in March 1980, only to be defeated in the full House. I used the time between the two sessions to do the research for this book, which began merely as a follow-up to the assertions made by NSP's lobbyists.

One of the reasons I wanted to pursue some of the issues that had been raised during the deliberations on HF 378 was that NSP had massive resources to disseminate whatever information they chose. If I was going to be able to compete with their lobbyists in the future, I had to have a command of the subject.

I used neutral sources as much as possible, rather than extremely antinuclear or pronuclear articles and books. Useful information can be found, however, in such antinuclear books as The Menace of Atomic Energy by Ralph Nader and John Abbotts and Poisoned Power: The Case against Nuclear Power Plants by John Gofman and Arthur Tamplin. Similarly, ardently pronuclear books, such as Petr Beckmann's The Health Hazards of NOT Going Nuclear, are provocative. I found, however, that the more I read, the less neutral I became on the subject of nuclear power. By the time the legislature reconvened in January 1980, I had become deeply antinuclear. I hope that the justification for that perspective will become clear in this book.

The theme of "hidden power" emerged as I researched the various facets of nuclear power. For one thing, on the physical level the atomic fission that unleashes tremendous heat and energy is never seen. Unlike a coal-fired facility, there is no way an observer can peek into the massive furnace that powers the nuclear plant. Enormously high levels of radiation in the reactor core make it virtually inaccessible even after the atomic reaction has concluded.

No technology with a commercial application owes a greater debt to federal subsidies than does nuclear power, and in that sense the research and development costs of nuclear power were hidden from ratepayers. Some of the costs of nuclear power have been hidden because as yet they have not been met. The cost of nuclear-waste disposal and decommissioning are prime examples. The federal government's willingness to provide liability insurance to utilities with nuclear power plants represents a hidden cost. Indeed, even advocates of nuclear power concede that if the owners of reactors had to get their insurance solely in the private marketplace, the cost to the ratepayer of that form of energy would go up significantly. Other hidden costs abound.

In addition to hidden costs, nuclear power poses serious, hidden health hazards. Radiation, which is odorless and colorless, can do lasting and even lethal damage to living organisms. It is incontestable that high levels of radiation cause death and serious disease; what is less easy to prove is the damage done by exposure on a long-term basis to low levels of radiation. Like leaders of the tobacco industry in the past, nuclear power advocates argue that because one cannot absolutely prove a dose-response relationship, society should assume that low-level radiation is safe until proved otherwise. However, by its very nature, the damage done by radiation may not show up for several generations.

For a number of reasons, the true alternatives to nuclear power have not been adequately publicized and promoted. In that sense, safer and less-expensive ways of meeting the nation's electrical needs have been hidden. The combination of the federal government, which developed and promoted nuclear energy, and some large corporations, which have strong economic interests in perpetuating it, has caused the alternatives to nuclear power to be obscured. Those who advocate a nonnuclear future and propose realistic and specific ways of achieving it lack the vast financial resources that are available to those with a contrary view. For that reason, the public is largely denied an exposure to the true choices that are available to be made. Hence, the options of industrial cogeneration of electricity, photovoltaic electricity, and organized electrical conservation, among many others, lack public emphasis and are in that sense hidden.

Several important issues are ignored in this book: among them, the questions of the morality of nuclear power and its obvious vulnerability to sabotage. It is clear that serious moral and ethical problems are involved in deciding on the future of this technology. Should corporate and government leaders have the right to make decisions on the construction and location of nuclear power plants without adequate consultation with and agreement by the local citizenry? Is there no obligation to inform women of child-bearing years in the affected area of the potential impact on them and their children of even low-level exposure to radiation in the event of an accident? When other, attainable alternatives are available today, is it right for our society to generate thousands of tons of highly radioactive waste that could be dangerous to people hundreds of centuries into the future? Should the legacy of our civilization to future generations include massive amounts of nuclear waste?

What safeguards ensure that large urban populations will not be threatened by potential saboteurs of nuclear plants? As the number of reactors grows, what additional security measures will be needed, and at what potential cost to the citizenry in the area of civil liberties?

This book does not attempt to answer those questions, profoundly important though they may be. Rather, it is my purpose merely to examine the following simple statement, often declared by nuclear power advocates: "Nuclear power is cheap, safe, and necessary."

1

A TRIP TO MONTICELLO AND BEYOND

February in Minnesota makes a person conscious of the need for energy. It can get so cold here in the winter that you do not really care where you get the heat, just that you get it, appreciate it, and hope that the system that delivers it to you will not fail. There is a kind of pride at being able to endure the cold when living in this state and a kind of fellowship the weather creates among the people here. Some of that sense of camaraderie was with me as I boarded the bus used by the House of Representatives' Energy Committee to travel to Northern States Power (NSP) Monticello nuclear power plant for a tour. I felt a sense of companionship but also a sense of anticipation, as well; I had just introduced legislation providing that no new nuclear power plants could be built in Minnesota until a means of disposing of the waste had been found. The trip to Monticello, a small town in central Minnesota, would give me a chance to be in the "lion's den" as well as a chance to get to know the other members of the Energy Committee better. We were evenly divided on partisan lines in the House, so I knew I would need at least one Republican vote to get the bill out of committee.

The thing I wanted most on that trip to the nuclear power plant was information. I had introduced my bill with little knowledge about about nuclear power and I wanted to learn as much as I could as quickly as I could from whatever source available. I did not want to embarrass myself in trying to pass the first major bill I was to carry as a legislator. Although I knew my bill might be viewed as antinuclear, my personal feelings were that nuclear power was cheap and probably necessary, although a little dangerous.

It was exciting to be in a nuclear power plant. It was massive and gave off a never-ceasing din of the sound of power, power that our guide told us was enough to provide electricity for hundreds of thousands of people in the Twin Cities, 40 or 50 miles away. What I remember most was staring into the perfectly clear pool of water containing the spent fuel rods, those elements of used-up uranium that had been in the reactor but no longer contained enough fissionable material to be of value inside that mysterious, unviewable structure. The pool of water was so clear you could see to the very bottom. The long fuel elements, stretched out in perfect, stationary rows, were made more impressive by thinking about the power they had generated earlier just a hundred feet away. I must admit that this nuclear waste was a little awesome to behold. What to do with that waste would be the subject of much of the early research for my bill. I was interested to hear what NSP's chief nuclear expert had to say about that.

After our tour we assembled in a spartan meeting room, our hard hats off, and sat in cheap, plastic chairs to drink coffee, eat rolls, and listen to NSP's story of nuclear power. I believe I was the only legislator to take notes. The NSP spokesman was a man in his fifties who had served 30 years in the United States Navy, most of which he had worked in the submarine service with the nuclear submarine and Polaris missile programs. He was the classic "nuclear priest" with a master's degree in mechanical engineering. As NSP's manager of nuclear-plant projects, he projected an image of credibility born of long experience and cool intelligence. His voice was deep and sure; his bearing was confident. There was no nonsense and no ambivalence in this man regarding the need for nuclear power. He said many things about the subject, but I focused on what he had to say regarding nuclear waste. The following three statements paraphrase what he said.

1. The volume of high-level waste is so small as to be practically inconsequential; the military generates 99 times more high-level waste than does commercial nuclear power.

He had worked many years for the military, so I assumed he knew what he was talking about. I felt a little foolish being so concerned about commercial nuclear waste when its volume was miniscule. Still, it was very radioactive waste and there was no agreed-upon method of disposing of it, so I felt justified in sponsoring the kind of legislation that I had just introduced.

2. The problems of the disposal of highly radioactive waste are political, not technical, in nature.

Experts like this NSP spokesman apparently had a pretty clear idea of what to do with the waste (although he did not really specify what it was), but for political reasons they were not able to dispose of it. As a politician, his statement made me feel defensive, but he exuded such dispassionate credibility that I figured he was probably more right than wrong. I would check into that later.

3. It is possible to calculate the cost of disposing of highly radioactive waste, and ratepayers are already paying for it in their monthly electric bills if they receive nuclear-generated electricity.

If the experts had a pretty good idea of what to do with the waste, it made sense that they could put a dollar figure on the eventual cost of disposing of it. I honestly had not realized that it was already being figured into my utility rates. This did not really bear on the validity of my proposed bill, but it was interesting to know and gave me the impression that the utility company was more on top of the waste-disposal problem than I had thought.

I left the Monticello nuclear power plant better informed on nuclear waste, or so I thought at the time. I figured many of the same statements would be made if NSP gave testimony against my bill, so I decided to do some follow-up research on those three statements. Much of that research went on during the legislative session, but some of it went on into the following summer after I had decided that there was another nuclear story that needed telling. My experience at Monticello and the subsequent knowledge I acquired started me down a path of inquiry that was fascinating, disturbing, and hopeful. Let me begin the other side of the story by telling you what I have learned about the three statements made by that NSP expert.

ON THE LOW VOLUME OF
COMMERCIAL HIGH-LEVEL WASTE

High-level waste does not mean what one would think it means, namely the spent fuel assemblies of the kind I had seen in the pool next to the reactor at Monticello. Indeed, the statutory definition of high-level waste is very narrow, and the use of the phrase is therefore most misleading. High-level waste is the watery by-product of a process called "reprocessing," a chemical means by which fissioned, or used-up, uranium and plutonium are removed from used fuel rods for use again. Because of the security risks of plutonium, the critical ingredient in a nuclear bomb, President Carter banned commercial reprocessing in April 1977, much to the disgust of the nuclear power industry. Hence, for the past few years there has been little growth in commercial high-level waste. At the same time the volume of spent fuel rods has grown apace. According to an NSP spokesman with whom I talked, each reactor generates from 25 to 40 tons of spent fuel elements each year. Even though they are not technically high-level waste, that spent fuel remains extremely radioactive after it has been removed from the reactor.

Meanwhile, the Defense Department has been generating high-level waste since World War II; precisely how much cannot be determined for security reasons. Hence, the comparison of commercial high-level waste, which has been generated only since the 1960s, with defense-generated high-level waste is inappropriate and very misleading.

Going beyond the semantic issue of what constitutes high-level waste, another important issue to consider in assessing the volume of commercial nuclear waste is its comparative radiotoxicity, or how radioactively toxic it is. Two Princeton University scholars, Hartmut Krugman and Frank von Hippel, have made the point that while the volume of military high-level waste is considerably greater than that of commercial power, the measurement of volume only is highly misleading.

> When account is taken of the fact that existing military high-level radioactive wastes are almost 100 times more dilute than projected high-level radioactive wastes from the reprocessing of fuel used in nuclear power plants, one finds that the present activities of fission products that have been generated by the still relatively small U.S. commercial nuclear energy program may already be approximately equal to those that have been generated by military programs.[1]

Krugman and von Hippel follow the Nuclear Regulatory Commission's (NCR) method of comparing the hazards of different kinds of fission by-products, using the total volume of water that would be needed to dilute the radioactive wastes to conform to limits set up by the NRC and specified in the Code of Federal Regulations. Be-

FIGURE 1.1

Estimated Civilian and Military High-Level Waste Inventories Measured in Terms of Their Strontium 90 Content as a Function of Time

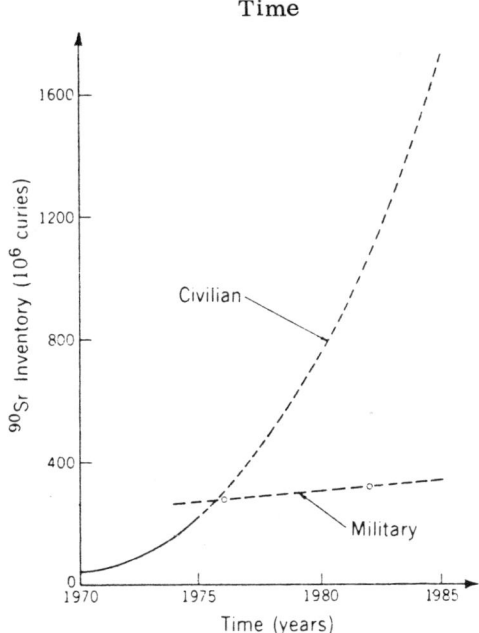

Source: Reprinted, by permission, from Hartmut Krugman and Frank von Hippel, "Radioactive Wastes: A Comparison of U.S. Military and Civilian Inventories," Science, August 26, 1977.

cause the hazard potential for the first few hundred years after fission is dominated by the fission by-product strontium 90, they use that as their basis of comparison. According to the two researchers, the military and civilian inventories of Sr-90 are about equal, and by 1985 the total civilian inventory will be increasing annually by an amount of the same order as that of the total military inventory. Figure 1.1 demonstrates clearly the rapid growth of quantities of Sr-90.

Strontium 90 is particularly hazardous biologically because it emits extremely penetrating radiation and large amounts of heat. More significantly, Sr-90 is chemically like calcium. This means that our bodies cannot distinguish between calcium and Sr-90 when building bone, and so Sr-90 stays with a person much longer than if it merely passed through one's system without attaching itself. The longer it stays in the body, the higher the probability of Sr-90 decaying radioactively and damaging bone cells or bone marrow, thus causing leukemia. It is generally agreed that this particularly dangerous by-product of nuclear fission requires isolation from the environment for a minimum of 1,000 years.

I was struck by the rate of increase of radioactively toxic waste as measured by the presence of Sr-90 that Krugman and von Hippel had uncovered. Using radiotoxicity as the key indicator, rather than just volume, the data of the two Princeton researchers show an alarming growth of nuclear waste. Figure 1.1 is most eloquent on this point.

Strontium 90 is just one of many fission waste products of a potentially lethal nature. Two University of Minnesota scientists, Donald P. Geesaman and Dean A. Abrahamson make a forceful point about the dangers of cesium 137.

> Consider the ton of fission wastes produced during a year of operation by a 1,000-megawatt (electric) nuclear power plant. Of this total inventory, ignore all but one fission product isotope, cesium-137. It is sufficiently unpleasant that if all the other fission products appeared as common ash, cesium-137 would itself constitute a major problem of radioactive waste. About 60 pounds or 3 percent of the reactor's yearly fission products appear as this waste material.
>
> As reference values, the official guidance for a maximum permissible body burden is about one trillionth of that amount, and local surface contamination of one thousandth of that amount per square mile would give an ambient exposure to ionizing radiation of about 1,000 times the usual natural background of exposure. The half-life for this isotope is 30 years, and hence over a social time interval, such as a generation, the radiological activity of cesium-137 wastes would diminish by only a half. The 60 pounds of cesium-137 produced by operating a nuclear power plant for a year have the potential of excluding humans from hundreds of square miles for decades. It is projected that the accumulated activities of cesium-137 in the United States during the 1980s will represent thousands of reactor-years of op-

eration. Fuel reprocessing plants, such as the one under construction at Barnwell, South Carolina, might have in localized surface storage the inventory produced by 50 such reactors during 10 years of operation, that is, almost 500 times the annual contribution of one typical reactor.

Cesium-137 is representative of the significant social characteristics associated with fission wastes; the amounts are physically small, the times involved may be politically long, and the potential for damage is very large. Society can be protected only if the material flows are totally contained, and the isolation of the ultimate waste is permanently guaranteed.[2]

Bearing in mind that the definition of <u>high-level waste</u> is narrow and that looking only at volume of waste is very misleading, we can better understand figures disseminated by the nuclear power industry. The amount of commercial waste indicated in the March 1979 report of the Interagency Review Group on Nuclear Waste Management is sobering: 80,000 cubic feet of high-level waste, 123 kilograms of transuranic waste (such as plutonium), and 2,300 metric tons of heavy metal in the form of the spent fuel discharged from commercial reactors. (The report also notes that commercial nuclear power caused 15.8 million cubic feet of low-level waste and that there were 140 million tons of mill tailings.) The simple fact is that there is a great deal of commercial nuclear waste and its volume is growing daily. Each reactor produces between 25 and 40 tons of spent fuel elements each year, according to an NSP spokesman I talked to. Commercial high-level nuclear-waste radiotoxicity exceeds that of military high-level waste, and the rate of growth of commercial nuclear waste greatly exceeds that of militarily generated waste.

ON WHETHER THE DISPOSAL PROBLEM
IS MAINLY POLITICAL, NOT TECHNICAL

When utility lobbyists say this problem is political and not technical, their implication is clear: if politicians did not slavishly follow the hysterical fluctuations in public sentiment on nuclear power safety, technicians would be able to go about their business of efficiently disposing of nuclear waste. Yet, as I will show later, any kind of research into the waste-disposal issue reveals that there are serious technical problems that still need to be resolved.

It should be granted at the outset that there are some serious political problems and institutional issues that will have to be resolved before a permanent repository for highly radioactive wastes is found. None is more important than that of the right of states to determine whether or not they will allow such waste disposal within their boundaries.

So far, political influence has given states de facto veto power on waste-disposal site selection. Kansas, Michigan, and New Mexico have all resisted overtures from the federal government to develop permanent repositories for highly radioactive nuclear waste within their boundaries. The federal government has consistently acquiesced to the states, although the constitutional case for states' rights in this area is debatable. An article by Patricia Lucas is important, since it shows how, in a future case, several provisions of the U.S. Constitution might cause the courts to find in favor of the federal government.

1. The Constitution is clear that on federal lands the federal government has exclusive jurisdiction; therefore, placing a repository on federal lands, in one of the western states for example, could be difficult to oppose by any given state.

2. The Constitution ensures federal preeminence in war powers and interstate commerce, both of which come into play in the regulation of nuclear power. In this case, that power could be used to site and build nuclear-waste repositories.

3. In Ray v. Atlantic Richfield Co., the Supreme Court of the United States affirmed that a state law is not valid if "Congress has either explicitly or implicitly declared that the states are prohibited from regulating" that area of concern. A strong case can be made that Congress does intend to preempt states' rights on the waste-disposal issue by virtue of the fact that federal regulations require permanent waste repositories to be located on federal property.[3]

However, no matter how the Constitution is interpreted on this matter, the states still represent a formidable political force. The political power of a state to resist attempts by the federal government to put nuclear waste within its boundaries is vividly illustrated by New Mexico and the proposed federal Waste Isolation Pilot Plant (WIPP). For the past several years the Department of Energy (DOE) has sought to demonstrate that highly radioactive waste could be disposed of safely, in this case in the thick beds of salt in the area around Carlsbad, New Mexico. The proposed WIPP was originally planned to be of such size as to be able to accommodate all of the nation's high-level military waste and used commercial fuel through the year 2000. Yet, after spending $87 million for the Carlsbad project, Congress rejected the idea of the DOE's proposed storage of 1,020 discarded fuel assemblies on the grounds that it did not want to mix military and commercial waste. More significant was the New Mexico governor's insistence that his state keep a veto power and be given licensing power if such a facility were constructed. New Mexico's resistance to providing a waste-disposal site was particularly significant because that state has a large economic interest in nuclear power; New Mexico is the source of almost half of the uranium used in nuclear-powered reactors. Such federal and state resistance to WIPP helped cause President Carter to finally cancel the project in February 1980.

To date, the political clout of states has been sufficient to bar the establishment of a permanent repository for highly radioactive waste, but as that waste continues to accumulate, the pressure on the federal government to put it somewhere could reach the point where the views of states will no longer prevail.

Two other political problems in nuclear-waste disposal have been neglect of the issue until recently and departmental fragmentation. From the time of President Eisenhower's "Atoms for Peace" in 1954 until 1974, very little was done to deal with the waste-disposal issue, and, although the nuclear power industry did little to force the issue, political leaders during that period must bear the ultimate blame for failure to attack the problem sooner. The lack of a sense of urgency may well have been understandable, but the result of neglect and delay has left us with a large technological problem and little time to solve it. There is even uncertainty as to who is clearly responsible for solving it.

Until 1974 the Atomic Energy Agency was in charge of almost every aspect of nuclear power: research, development, promotion, and regulation. The obvious conflict of interest inherent in having the same agency promote and regulate nuclear power prompted Congress to pass the Energy Reorganization Act in 1974, thus changing waste-disposal governance. Congress created the Nuclear Regulatory Commission (NRC) and gave it the regulatory and licensing power of the old Atomic Energy Agency, as well as many of the latter's personnel. The same act gave the Energy Research and Development Administration (ERDA) atomic research and development responsibilities, which were shifted to the Department of Energy when that was created in 1977. Currently, DOE is responsible for developing and demonstrating waste-disposal technology; the NRC is charged with assessing risks and establishing standards for all phases of the nuclear-fuel cycle and has appropriate licensing powers; and the Environmental Protection Agency (EPA) establishes general environmental standards for waste-management activities, focusing mainly on permissible levels of various forms and kinds of radiation.

As well, the U.S. Department of Transportation must formulate and enforce safety standards for the transportation of radioactive waste and deal with routing issues raised by the increasing number of state and local bodies opposing the transportation of such waste through their jurisdiction. The Department of the Interior, through the U.S. Geological Survey, must conduct appropriate tests to arrive at a preferred site for permanent waste disposal.

Because some of these agencies are relatively new and the impetus to solve the nuclear-waste-disposal problem has developed only recently, each respective agency is really in the formative stages of learning how to do its job. For each agency to do its job correctly, coordination with the other relevant agencies is essential. For example, it is necessary for the DOE to have a sense of the NRC's probable licensing requirements in its development of a technology for waste disposal, just as it is important for the NRC to have a clear idea of what the EPA's radiation standards are likely to be.

An attempt to overcome such fragmentation and to improve coordination took the form of the Interagency Review Group on Nuclear Waste Management established by President Carter. Representatives from 14 agencies served on the Interagency Review Group (IRG), and their report to the president contained a series of recommendations ranging from criteria to be used in assessing a permanent waste-disposal system to improving state-federal relations on this subject. Nonetheless, fragmentation of effort persists as a problem in nuclear-waste disposal, which is a result of political decision and indecision.

To his credit, on February 12, 1980, President Carter presented the most comprehensive and ambitious nuclear-waste-disposal management plan to date. The $670 million program addresses the need to dispose of all kinds of radioactive waste, including that which is highly radioactive. In order to ensure better cooperation with the states, Carter established the State Planning Council on Radioactive Waste Management, which is made up of 19 persons, 15 of whom represent individual states. The council is to advise the executive branch on the siting, construction, and desired operation of any high-level waste facility. After adequate research into the best way to package the waste and the least risky geological repository for it (a process that the government hopes will take only five years), the government would begin constructing an actual waste-disposal facility to be ready by 1995. As will soon become evident, precise prediction of the timing of this task is difficult in view of the numerous technical issues that are still unresolved.

Carter also proposed three test facilities that would allow the waste to be retrieved: a granite facility in Nevada that has already accepted 11 cannisters, each containing 200 rods of spent fuel; a basalt facility in the state of Washington; and a salt facility at a site to be determined after preliminary research in Utah, Texas, Louisiana, and Mississippi.

To acknowledge that there has been a long tradition of poor political leadership on this issue and that certain institutional fragmentation grew out of political decisions does not lessen the seriousness of the technical problems that are still unsolved. Moreover, until there is a clear, technical resolution of the problem, the political opposition from candidate states to the building of long-term repositories within their boundaries will persist.

The challenge of developing a safe and permanent method of disposing of highly radioactive waste is unprecedented in human history. Never before has it been necessary to devise a technology that would have to work as well thousands of years into the future as it does today. To be able to plan for such a long period is necessary because some by-products of nuclear fission, such as plutonium, have half-lives of over 20,000 years. Add to that the fact that the effort to research and develop long-term nuclear-waste-disposal technology has begun only recently. One is struck both with the vastness of the task and how little has been done so far. Because thousands of tons of highly radioactive waste exist right now, there is no alternative but to figure out some means of isolating it; most people would agree that corners should not be cut in doing so. The higher our standards of safety are for a nuclear-waste-disposal system, the more research that will have to be done. If standards were sufficiently low, we could dispose of long-lived, highly radioactive waste at once; in that sense, waste-disposal technology is already available. On the other hand, if the nation is serious about finding the very best possible means of doing the job, it will take time. Moreover, as we shall see, the nature of the studies, research, and investigations that are needed are such that it is virtually impossible to estimate when enough will be known to proceed safely.

Let us assume that the goal of long-term, nuclear-waste disposal is to ensure its safe and permanent isolation from human beings. The question then becomes, What is the best repository for long-lived radioactive waste? In the past, shooting the waste into space was considered a possible means of disposing of it, but there is now a strong consensus that a deep geological repository is the best way to dispose of nuclear waste for the near future. Beyond agreement on that, however, there are few points on which there is scientific consensus.

There is not even agreement on what specific form the waste should take. While it is generally agreed that the waste should be solidified, there are a number of variables that will determine the ideal form, including the age of the waste, the transportation involved, the geological medium chosen for constructing a repository, and the overall waste-disposal system agreed upon. When those issues are resolved, the solid form could be one of a number of possibilities including glass, ceramics, cement and concrete composites, and metalmatrix composites, among others.

More important even than the preferred solidified form is the determination of the best geological home for the highly radioactive waste. Should it be placed under the sea, or is a continental repository preferable? Although the U.S. Congress banned all dumping of nuclear waste in the sea in 1972, a multinational effort called the Seabed Disposal Program is studying the technical and environmental feasibility of burying high-level waste 30 to 80 meters below the abyssal depths of the sea, depths of more than 15,000 feet. Some participants in the program hope to begin placing cannisters of high-level waste below the ocean floor by 1990-95. As well, the DOE spent nearly $6 million in 1980 (increasing to $15 million by 1984) to study seabed waste disposal.

A possible means of such disposal would be to put a spent-fuel assembly in one-foot-wide, twelve-foot-long missile-shaped cannisters that would be lowered by cable and then dropped so that they would reach speeds in excess of 100 miles per hour, thus burying themselves in the bottom sediments. Serious scientific and environmental questions remain to be answered before a decision on seabed radioactive-waste disposal is made. One drawback of placing the waste in the ocean can be finding it later. Of 47,500 barrels of low-level radioactive waste dumped off the coast of San Francisco in the 1950s and 1960s, only 150 have been found, according to an EPA official in that area. A continental site, while fraught with political problems, allows easier demarcation (monitoring the possible retrieval) than does one under the sea.

The great advantage of a seabed disposal site is that it largely circumvents constituents, the biggest obstacle to quick and easy nuclear-waste disposal. Far fewer citizens today would be inclined to raise a ruckus if nuclear waste is simply placed in the sea rather than on land. It certainly would be in line with the manner in which nuclear waste has always been treated: ignore it or hide it and let the future take care of itself.

From a technical point of view, the permanent waste-disposal question is primarily a geological and geophysical one. Earth scientists must work to find a repository site for the waste that meets these criteria: free of seismic activity such as earthquakes; unlikely to be permeated by water that could transport the radioactive material beyond the repository; in rocks thick enough to protect the waste-storage facility from disturbances on the earth's surface above it and of such chemical qualities as to be able to resist the heat of the waste or its containers; and in a location unlikely to be intruded upon in the future by those engaged in mineral exploration. Moreover, it should be remembered that these criteria must be met for thousands of years.

A look at some of the studies to be done by the United States Geological Survey (USGS) illustrated the magnitude and complexity of the research still necessary before nuclear-waste disposal can be considered solved. George D. DeBuchannane, chief of the Office of Radiohydrology, Water Resource Division, USGS, told the Minnesota House Energy Committee on April 10, 1979, that a number of studies still need to be conducted before the nuclear-waste-disposal puzzle is solved. The following are merely some of the needed studies:

1. A study is needed to identify environments with effective multiple barriers to the transport of radioactive material. This includes a reconnaissance study of regions throughout the nation to see if they meet specified geochemical and other criteria.

2. Based on the finding of multiple-barrier environments, the specific rock formations that qualify must undergo careful analysis. This will include a look at western shales, granite and related crystalline rocks, salt, and other possible kinds of rock.

3. Through the use of geophysics and geochronology, studies have to be made concerning the likelihood of tilting, bulging, faulting, erosion, or other changes that could affect the suitability of a repository over time.

4. The permeability of fractured rocks and the behavior of water and vapors under various conditions must be examined, particularly as they relate to the presence of highly radioactive waste in a given geological environment.

5. Ways must be developed to adequately ensure long-term prediction of geological events and changes such as earthquakes, faults, and climatic shifts that could alter the ability of the repository to hold the waste. It should be remembered that some highly radioactive waste needs to be isolated for periods of extreme duration.

6. Models need to be established to assess the movement of radioactive materials in different kinds of rocks to help determine those most resistant to such movement. The findings of studies using such models will have much to do with the final selection by the Department of Energy of the most appropriate radioactive-waste-disposal site.[4]

Essentially, the earth scientists must determine the best geological repository for different kinds of waste and, after exhaustive research, reach a judgment on where it should go, in part based on risk-assessment models developed in cooperation with the DOE and the NRC. The final recommended site will not be absolutely risk-free, but it will be as close as is humanly possible. Obviously, over the course of thousands of years, geological and other conditions could change in such a way as to expose future human beings to considerable radiation from nuclear waste.

Beyond the technical problems to be addressed by geologists, geochemists, and geophysicists, there are two other kinds of technical problems of considerable importance: how safely to transport large quantities of nuclear waste and how to ensure adequate security to avert sabotage or theft. Because a great majority of the spent fuel rods are simply kept in large storage pools next to the reactors in which they were used, there has been little need to develop and refine means of transporting the great volume of nuclear waste that exists. To be sure, one can simply put the spent fuel rods in containers, load them on trucks, and take them to their destination, as has been done on a limited basis in cases of reprocessing. However, when there are many such trucks traveling the roads of the United States, rigorous standards for safety will have to be established and met, neither of which has yet occurred. There has been some testing of casks used to carry high-level waste, but much remains to be done in regulating the safe handling of highly radioactive waste in its transportation.

Much also needs to be done in the area of avoiding sabotage or theft of this potentially lethal material. In the case of plutonium, which is used in nuclear bombs, it is not a far-fetched scenario to envision a criminal or political zealot taking advantage of its availability. Some kind of technology must be developed to minimize that risk as well.

The technical feasibility of nuclear-waste disposal, in my view, boils down to how careful we want to be and how high our standards should be in determining the best site, and means of, isolating that waste. If our standards remain high, it is clear that much more research needs to be done. Moreover, as with any kind of research, it is unrealistic to predict at what date the research will have solved the problem. In view of that fact, pell-mell development of new nuclear power plants seems to be very unwise.

ON THE COST OF DISPOSING OF
HIGHLY RADIOACTIVE WASTE

It seems self-evident that until there is agreement on the means of disposing of highly radioactive waste, there can be no meaningful basis for estimating the cost of such disposal. To make such an estimate would be comparable to calculating the cost of building a house without knowing its specifications. That it is illogical to do so has not inhibited the government from estimating the cost of nuclear-waste disposal, which in turn has stimulated those opposed to nuclear power to go through the same exercise. The federal government estimates that over the next 20 years the total cost of the radioactive-waste-management program will be $30 billion.[5] Such estimates are also expressed in cents per kilowatt-hour of electricity: the range of total electrical costs is generally from 3 cents to 12 cents/kwh in the United States. The U.S. Department of Energy estimates the cost of waste disposal to be about 0.1 cent/kwh, or a 25-fold increase since the Atomic Energy Commission's 1971 estimate.[6] On the other hand, MHB Technical Associates, a nuclear engineering consulting firm, estimated a cost range from 0.1 cent/kwh to 2.0 cents/kwh in a 1978 study it did for the Natural Resources Defense Council. According to the MHB study, the factors most likely to cause great cost increases related to the same kinds of management blunders and institutional problems that have characterized the waste-disposal program in the past.

Perhaps the most extreme example of underestimating the cost of nuclear-waste disposal occurred at the reprocessing facility in West Valley, New York, which operated from 1966 to 1972. The firm operating the facility, Nuclear Fuel Services (NFS), originally set aside $4 million for waste management, but an estimate made by the NRC in 1976 ranged up to $540 million[7] and the most recent DOE figures range up to $1.2 billion.[8] It is as likely as not that even ignoring the unusual West Valley case, the cost of waste disposal has been underestimated by such advocates of nuclear power as the federal government.

The West Valley case throws into question not only cost estimates of any kind for nuclear-waste disposal but the technical competence of the industry as a whole. When an estimate can jump from $4 million to $1.2 billion, one is left wondering if the nuclear power system as a whole is as tight and competently managed as one is led to believe in the expensive advertisements run by utilities. The chapter dealing with reactor accidents only reinforces this concern. Moreover, as will be shown later, the higher our standards of safety for nuclear-waste disposal become, the more expensive it will become to perform that task. Regardless of those standards and the resulting rise in costs, it is clear to me that specific estimates of the cost for nuclear-waste disposal in the absence of an agreed-upon technology is a silly and misleading exercise. What is certain is that today's and tomorrow's taxpayers and ratepayers will be footing a bill of unknown magnitude for decisions made today by utilities that choose to build new nuclear power plants.

Another important part of the total cost of nuclear-waste management and disposal relates to the cost of closing down and isolating heavily contaminated nuclear power plants, as well as all other facilities needed in making usable nuclear fuel. That process is called "decommissioning," a significant capital expense that occurs after the 30- to 35-year life of a nuclear power plant. Unlike most other kinds of facilities, which have some salvage value after they have run their useful lives, nuclear plants have what is called "negative salvage value." That means that the disposing of the facility presents only an additional cost with no off-setting value affixed to the highly contaminated property.

Decommissioning can entail immediate dismantling of a facility or a delayed dismantling 30 years or even 100 years after it has ceased operating. The NRC is in the process of establishing decommissioning regulations for each kind of facility requiring it, from uranium-milling plants to low-level-waste burial grounds.

As with spent-fuel and other nuclear-waste disposal, there is no agreed-upon technology for the best way to decommission a nuclear facility; dismantling may be best, or entombing for a long period may make more sense, depending on what kind of facility is in question. Once again I was struck with the fact that such a basic insue remains unresolved after 72 reactors have been put into operation and another 90 are on the way. Moreover, reactors are only one kind of facility that must be decommissioned.

The cost of decommissioning is still a matter of speculation. A study done for the NRC by Batelle Pacific Northwest Laboratory in 1979 indicated that a 1,175-megawatt pressurized-water reactor would cost $33.3 million to dismantle immediately (in 1978 dollars).[9] On the other hand, in February 1977 an estimate made by NUS Corporation, an engineering consulting firm, indicated that the 436-megawatt San Onofre No. 1 power plant in California, also a pressurized-water reactor, would cost from $63 million to $78 million (1977 dollars) to decommission in a similar manner.[10] Any such cost will vary according to the site and design of the facility as well as the mode and timing of decommissioning.

Some utility companies maintain that both waste-disposal and decommissioning costs are figured into their rates. Once again, however, it seems puzzling to make such an assertion when the true cost of both activities is unknown. Beyond that, the cost of decommissioning all facilities associated with the manufacture of nuclear fuel surely is a cost that has yet to express itself in the cost of that fuel. When it does, that part of the cost of nuclear power will escalate even further.

CONCLUSIONS AND COMMENTS

Several points stood out as I was reading through the literature on the disposal of highly radioactive waste.

1. A serious and sustained effort to deal with this problem has begun only recently. There have been minor efforts in the past to study the issue, but the real impetus for action has occurred just in the past few years, in part because the pools containing spent fuel rods are getting more and more crowded.

It is nearly unbelievable to me that the government and the utilities allowed this technology to be initiated, promoted, and spread without the issue of waste disposal being resolved. Now a solution that must last literally thousands of years must be found virtually immediately.

2. Two factors have distorted the discussion of the volume of commercial nuclear waste: high-level waste has a very narrow statutory definition that does not even include the highly radioactive spent fuel rods; and discussion of volume alone, without an understanding of how radioactively toxic the substance is, tends to make commercial nuclear waste appear to be nearly insignificant compared with military nuclear waste, a serious misperception.

3. The volume of commercial waste is growing at a very fast pace. If the volume of radioactively toxic material is the standard of measurement, commercial nuclear waste is growing far faster than military nuclear waste.

4. Candidate states for repositories of highly radioactive waste have been consistent and effective in expressing their opposition to the siting of such a facility within their boundaries. There is little indication that such opposition will diminish.

5. Many basic technical problems remain in determining the best kind of geological repository for highly radioactive nuclear waste. Considerable research is needed before a consensus will be reached on the best site for such a repository, and estimating when that research can be finished is indeed speculative.

6. Equally speculative are estimates on the cost of highly radioactive nuclear-waste disposal, including the cost of decommissioning all the facilities involved in the nuclear-fuel cycle. It is likely that when those costs are truly accounted for, they will significantly increase the amount paid either by the ratepayer or taxpayer.

It is clear that my research led to very different facts and conclusions from those presented by the NSP spokesman at the Monticello nuclear power plant. The waste-disposal problem is far from solved from a technical point of view; for that reason, its costs cannot have been accounted for as yet. As I continued to push my bill, still in search of some Republican support, I decided to begin looking more closely at the economic arguments. It was obvious to me that the unsolved nuclear-waste problem constituted a hidden cost, and I began to wonder about the magnitude of other such costs not yet truly figured into the ratepayer's bill.

At Monticello the NSP spokesman had also said the costs of nuclear and coal power were a "wash," even disregarding the hidden costs. He said the capital costs of nuclear power had gone up so much in recent years that the fuel savings of nuclear versus coal no longer made atomic power economically superior. His comments caused me to look at the whole question of the cost of nuclear power more closely. I was surprised at what I was to learn.

NOTES

1. Hartmut Krugman and Frank von Hippel, "Radioactive Wastes: A Comparison of U.S. Military and Civilian Inventories," Science 197 (August 27, 1977): 883.

2. Dean E. Abrahamson and Donald P. Geesaman, "The Dilemma of Fission Power," Bulletin of Atomic Scientists, November 1974, p. 38.

3. Patricia Lucas, "Nuclear Waste Management: A Challenge to Federalism," Ecology Law Quarterly 7 (1979): 917-53.

4. Minnesota House of Representatives, House Energy Committee, Subcommittee on Utilities, Testimony of George D. DeBuchannane, April 10, 1979, at St. Paul, Minnesota.

5. U.S., Department of Energy, Division of Nuclear Waste Management Programs, The National Plan for Radioactive Waste Management, vol. 1: Introduction and Highlight, Working Draft 3, Rockville, Md., pp. 1-7.

6. U.S., Department of Energy, Preliminary Estimates of the Charge for Spent Fuel Storage and Disposal Service, Washington, D.C., July 1978.

7. Batelle Pacific Northwest, Alternative Processes for Managing Existing Commercial High-Level Radioactive Wastes, report prepared for the U.S. Nuclear Regulatory Commission, NUREG-0043, April 1976, pp. 141, 142.

8. U.S., Department of Energy, Western New York Nuclear Service Center Study: Final Report for Public Comment, TID-28905-1, November 1978.

9. Batelle Pacific Northwest, Summary of Preliminary Draft GEIS on Nuclear Facility Decommissioning, report prepared for the U.S. Nuclear Regulatory Commission, 1979.

10. Jon R. Stouky and E. J. Ricer, San Onofre Nuclear Generating Station Decommissioning Alternatives, report prepared by NUS Corporation for Southern California Edison, February 1977.

REFERENCES

Articles

Abbotts, John. "Radioactive Waste: A Technical Solution." Bulletin of Atomic Scientists, October 1979, pp. 12-17.

Abrahamson, Dean E., and Donald P. Geesaman. "The Dilemma of Fission Power." Bulletin of Atomic Scientists, November 1974, pp. 37, 41.

Burnham, David. "Growing Waste Problem Threatens Nuclear Future." New York Times, July 9, 1979, p. 1.

_____. "Wide Differences Persist over Safety and Methods of Nuclear Waste Disposal." New York Times, July 10, 1979, p. B-8.

Kerr, Richard A. "Nuclear Waste Disposal: Alternatives to Solidification in Glass Proposed." Science 204 (April 20, 1979): 289-91.

Krugman, Hartmut, and Frank von Hippel. "Radioactive Wastes: A Comparison of U.S. Military and Civilian Inventories." Science 197 (August 27, 1977): 883.

Lucas, Patricia. "Nuclear Waste Management: A Challenge to Federalism." Ecology Law Quarterly 7 (1979): 917-53.

Soloman, Norman. "Nuclear Burial Ground." Progressive, April 1979, pp. 33-39.

von Graevenitz, Alexander. "Nuclear Waste Disposal: Politics and Technology." Commonsense, Winter 1979, pp. 58-64.

Zinberg, Dorothy. "The Public and Nuclear Waste Management." Bulletin of Atomic Scientists, January 1979, pp. 34-39.

Reports

Batelle Pacific Northwest. Alternative Processes for Managing Existing Commercial High-Level Radioactive Wastes. Report prepared for U.S. Nuclear Regulatory Commission, NUREG-0043, April 1976.

_____. Summary of Preliminary Draft GEIS on Nuclear Facility Decommissioning. Report prepared for the U.S. Nuclear Regulatory Commission, 1979.

Carter, Jimmy. "President Carter's Statement of Nuclear Waste Management Policy." Nucleonics Week, vol. 21 (February 14, 1980).

MHB Technical Associates. Spent Fuel Disposal Costs. Report prepared for Natural Resources Defense Council, August 1978.

Pelham, Ann. "Underground Rock Storage Proposed for Nuclear Waste." Congressional Quarterly, February 16, 1980, p. 395.

Roisman, Anthony Z., and Thomas B. Cochran. Promoting Nuclear Power: Department of Energy Nuclear Waste Policy. Report prepared for Natural Resources Defense Council, January 1979.

Stouky, Jon R., and E. J. Ricer. San Onofre Nuclear Generating Station Decommissioning Alternatives. Report prepared by NUS Corporation for Southern California Edison, February 1977.

U.S., Department of Energy. Preliminary Estimates of the Charge for Spent Fuel Storage and Disposal Service. Washington, D.C., July 1978.

_____. President's Policy Statement on Comprehensive Radioactive Waste Management Program, State Planning Council Fact Sheet on Radioactive Waste Management Program. Washington, D.C.: Department of Energy, February 12, 1980.

_____. Western New York Nuclear Service Center Study: Final Report for Public Comment, TID-28905-1, November 1978.

U.S., Department of Energy, Division of Nuclear Waste Management Programs. The National Plan for Radioactive Waste Management. Vol. 1: Introduction and Highlight, Working Draft 3, Rockville, Md., pp. 1-7.

U.S., Interagency Review Group. <u>Report to the President by the Interagency Review Group on Nuclear Waste Management</u>. Washington, D.C.: Department of Energy, March 1979.

Speeches

Minnesota, House of Representatives, House Energy Committee, Subcommittee on Utilities. Testimony of George D. DeBuchannane, April 10, 1979, at St. Paul.

2

THE SURPRISING ECONOMICS OF NUCLEAR POWER

I had always believed that the main argument for nuclear power was its low cost. I had heard that it was once considered "too cheap to meter," and I believed that nuclear power benefited from economies of scale, low fuel costs, and the most modern and reliable technology. In the 1960s, when Minnesota's two nuclear power plants were built, those assumptions held the day. Northern States Power Company (NSP) generates over 40 percent of its power through nuclear energy, and over one-third of Minnesota's electricity is produced from nuclear power. The fact is that operating nuclear power plants do generate relatively inexpensive electricity, disregarding the kind of hidden costs we have already seen related to waste disposal and decommissioning. But current economics of nuclear power are very different.

As I began to research the economics of nuclear power, it became apparent that the picture had changed drastically in the 1970s. Capital costs of nuclear power went up at an average of nearly 20 percent a year in that decade. Design and regulatory requirements were to account for over half of those cost increases. By the end of the 1970s there were far more cancellations of orders for reactors than new orders, for reasons quite unrelated to the accident at Three Mile Island that occurred in 1979. Nuclear power was getting to be too expensive, and in the eyes of some, financially too risky to be as attractive as it once had been.

Interestingly, my research on the economics of nuclear power was greatly assisted by three men who are members of the private business community in Minneapolis: a venture capitalist, a corporate lawyer, and a funds manager who also advises a local brokerage house. Each had come to an antinuclear perspective from viewing the new economic realities of this energy form; none could be viewed as excessively cautious about nuclear power simply from a safety point of view. In my mind it is significant that each works in a highly competitive business where accurate and true cost accounting are rewarded. I felt lucky to have access to all three. The venture capitalist helped explain the key economic issues that must be analyzed when assessing nuclear power; the lawyer elucidated what will be presented later as a test case for the economics of nuclear power; and the funds manager showed me his work in comparing the stock market performance of heavily nuclear electrical utilities with those with little or no nuclear power. I also received cooperation from NSP staff in helping me understand their side of the story, particularly in the case of the test case of their proposed, but ultimately rejected, 1,100-megawatt nuclear facility near Durand, Wisconsin. (NSP was cooperative on technical explanations as well, and the definition of the "discount rate" that follows later is directly taken from that company.)

I started my inquiry into the economics of nuclear power asking the wrong question. I focused on the costs of nuclear power compared with coal, and it was only late in the summer of 1979 that I became aware of other, better alternatives that were also economically competitive or even superior. Still, discussion of nuclear versus coal provided a good reference point for understanding the changing economic realities of nuclear power. For that reason discussion of the test case of a proposed western Wisconsin nuclear facility is included.

The other principal element relating to cost that I wanted to learn about relates to the many hidden costs of nuclear power. Before getting into this project, I had given very little thought to the role of the federal government in the development and promotion of nuclear power. As you will see, that role is enormous. In fact, based on what I have read, I am convinced that without massive government assistance, nuclear power could never have come into existence or survived.

Commercial nuclear power is an unusual technology because its origins are found in government, not the private sector of the economy. From its very beginning there has been no clear demarcation between nuclear power as a public "product" and as a private-sector "product." The research and development of nuclear power were paid for by the taxpayer, and the current cost of using nuclear-power-generated electricity is borne by the ratepayer. The future cost of nuclear-waste disposal could be divided between the ratepayer and the taxpayer, as could the cost of decommissioning nuclear power plants to make them safe after their useful lives are completed. Whether ratepayer or taxpayer, however, the ultimate burden is borne by the citizen/consumer; and before nuclear power spreads further, the citizen/consumer should be aware of its costs, whether he pays it in monthly electrical bills or in federal income taxes.

Nuclear electricity is not sold in an open and competitive marketplace where the consumer is able to choose between different companies. Indeed, most consumers have virtually no power to express their wishes regarding how electricity is generated or how much each kilowatt-hour of it will cost. Nuclear electricity exists in a setting of regulated monopoly so that supply and demand do not determine its price strictly by market forces. Instead, the price is determined by what utility companies or other providers of electricity can persuade their regulators to permit them to charge for the product. Since the regulators—often called public service commissions—do not consider it in the public interest to jeopardize the financial well-being of the companies they regulate, requests for rate increases caused by higher costs or new power plants have traditionally been approved. The financial-investment community sees the regulatory climate as one of the most important factors in assessing any given public utility's stocks and bonds, so favorable rulings by public service commissions are critically important to corporate providers of electricity. Utility companies argue that the better treatment they get from regulators, the lower their cost of capital and the less expensive electricity will ultimately be.

The fact that nuclear electricity occurs in a setting of regulated monopoly is important for three reasons. The cost to the consumer is determined by regulators, not the marketplace. Lengthy hearings occur and complicated evidence is adduced in setting the price of electricity in a way that is beyond the comprehension and control of the average citizen/consumer. In that sense, the process that decides the amount of one's electrical bill is hidden and quite inaccessible to the consumer.

The incentive to cut costs, found in businesses that do not enjoy monopoly status, is partly missing in the case of a regulated monopoly. Managerial energies can shift away from concerted efforts to keep costs down and toward convincing the regulators of the need to increase rates in order to accommodate higher costs. While stockholders will not stand for gross excesses in the cost of running an investor-owned utility, the bottom line for a government-regulated utility is influenced more by revenues the regulators will decide to allow, a fact that blunts the usual business edge to cut costs. In fact, it is common to find automatic adjustment clauses in regulatory agreements, so that increases in the cost of fuel, for example, are passed along directly to the citizen/consumer without their having the benefit of a public hearing on the increases.

Unlike most other businesses, the monopolies that deliver electricity to consumers simply cannot be allowed to fail. Even if there are gross errors in judgment by the management of firms generating and selling electricity, the public cannot afford to have those firms go out of business. When there is a monopoly, by definition there is no place else to get the product. The result is that poor judgment on matters such as how much generating capacity to build, and at what likely cost, is not fully subject to marketplace discipline. Where other businesses might simply go out of existence, regulated monopolies can request and receive rate increases to compensate for their mistakes.

The lack of adequate incentives to keep costs down becomes clear and relevant when one looks at the economic trends of nuclear power during the 1970s.

RISING COST OF NUCLEAR POWER

The two critical components in the cost of nuclear power are capital (bricks and mortar and the cost of financing them) and fuel. The capital costs of nuclear power plants have increased at a considerably greater rate than that of inflation for the economy as a whole during the 1970s, and the cost of fuel has gone up sevenfold. Proposed new nuclear power plants promise to produce electricity that is considerably more expensive as a result of the new economic realities. Even disregarding the hidden costs of nuclear power, the trend of the past decade makes one wonder whether nuclear power is even affordable any longer.

Capital Costs

More than inflation, required changes in the design of nuclear power plants have caused capital costs to soar. Design changes are mandated to make nuclear power plants as safe as humanly possible, and the attempt to ensure even greater safety in the future will in-

FIGURE 2.1
Regulatory Growth of the Nuclear Industry

Source: Reprinted, by permission, from
Power Engineering, May 1977, p. 43.
© 1955 by Technical Publishing Co.

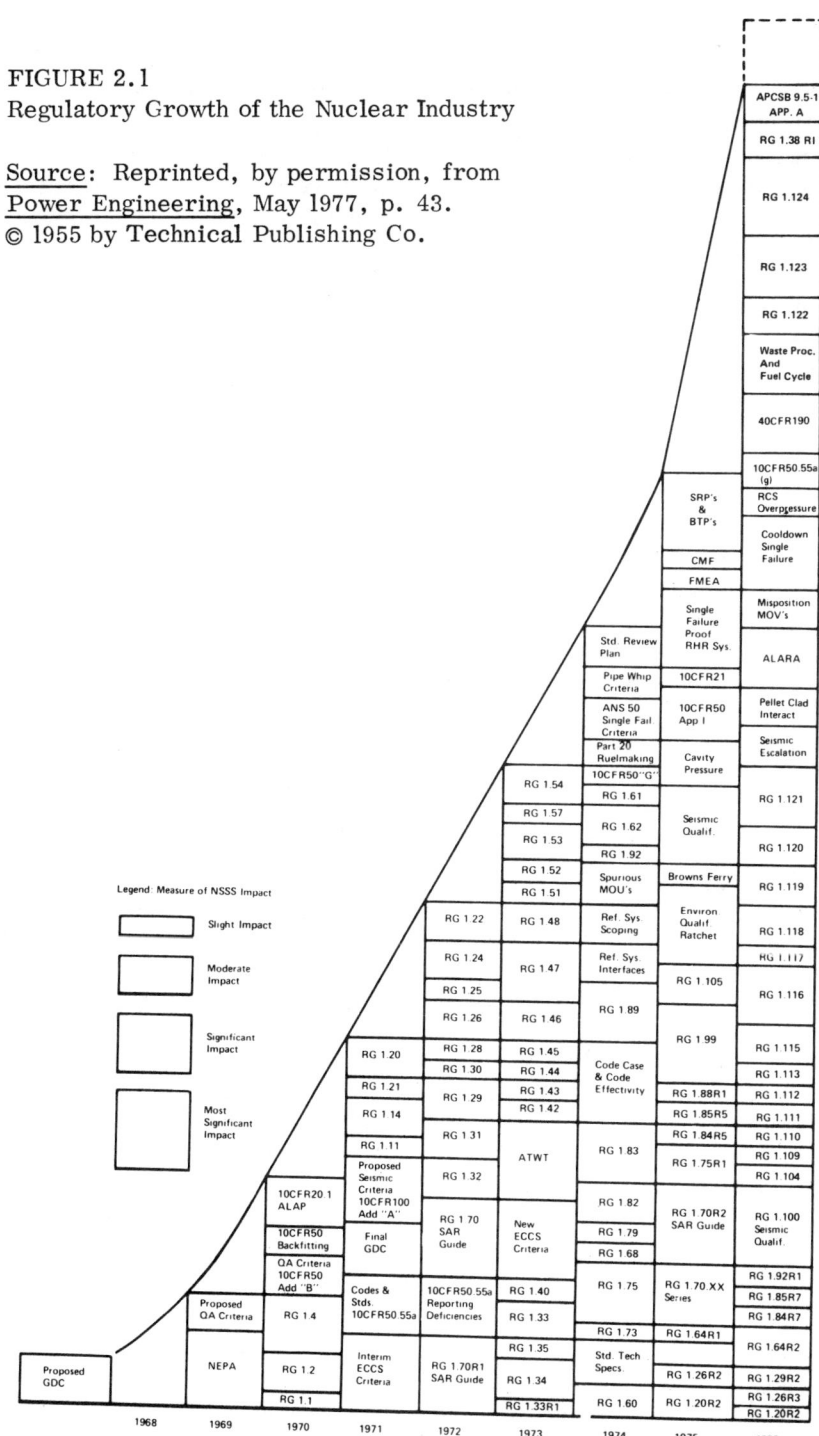

evitably lead to proportionally higher costs for nuclear power plants. Even before the accident at Three Mile Island, new regulatory and design requirements grew remarkably, as Figure 2.1 shows. While advocates of nuclear power have traditionally bristled at new regulations and maintained that they artificially raised the cost of nuclear power, it is unlikely that this trend will be reversed. Even before Three Mile Island, regulatory and design-cost increases added about 12 percent annually to the bill for nuclear electricity.

In addition to regulatory and design changes, the lead time for constructing nuclear power plants has been extended owing to tougher licensing requirements and increasing opposition from the affected communities. Whereas the lead time on a nuclear facility was five or six years at the beginning of the 1970s, the necessary time had nearly doubled by the end of the decade. In an inflationary period, longer lead times become very costly. Furthermore, construction costs rose about 10 percent a year during the 1970s. The more massive the facility, of course, the more construction-cost increases will affect the total cost of a project; and nuclear power plants are, by their very nature, the most capital-intensive means of generating electricity. Worse still, the cost of financing power-plant construction soared in the 1970s as this industry fell victim to high interest rates, as did the entire economy. Indeed, according to some analysts, interest costs led all others in the overall escalation of the cost of nuclear power in the past decade.

Large increases in capital costs have caused authorities in some states to cancel proposed nuclear power plants. The New York State Power Authority, for example, decided in early April 1979 to scrap their plans for the 1,200-megawatt Green County nuclear power plant. The authority cited the increase of $1.3 billion in capital costs over the original estimate of $1.8 billion during the period between June 1977 and early 1979. Similarly, Consumers Power in Michigan raised its estimate for a proposed two-unit nuclear plant from $1.67 billion to $3.1 billion in the early part of 1980. This jump in costs could mean a one-third increase to that firm's customers, according to a <u>Wall Street Journal</u> article of March 4, 1980.[1] The same article cited an estimate by a nuclear power economic expert, Irvin Bupp of Harvard, that the post-Three Mile Island furor over safety will add an additional two years to the completion time of each nuclear-plant project.

Reliability of Nuclear Power

It is important to assess how reliable nuclear power is when evaluating its true cost. If a car owner gets 40 miles to the gallon in a comfortable, well-handling, moderately priced automobile, that person should be satisfied. However, if the car only works half the time, the investment is greater than the performance, and the true cost of the vehicle is much higher than the sticker price, particularly if the alternative is a rented car or a taxi. In the case of electrical power plants, the measurement of reliability is called the capacity factor. The <u>capacity factor</u> is a measurement of the amount of electricity generated over the course of a year by a power plant compared with the potential electricity that could be generated if the system worked perfectly and continuously. It can be expressed in this equation:

$$\frac{\text{Actual kilowatt-hours per year}}{8,760 \text{ hours per year} \times \text{design power (in megawatts)}} \times 100 = \text{Capacity factor of a power plant}$$

Since the average nuclear power plant takes a month to refuel, which requires shutting the reactor down, the best possible capacity factor that could be expected would be a little over 90 percent.

The capacity factor has a strong effect on total capital costs. The utility has to recover the cost of constructing a plant by charging a portion of that cost to its customers each year, amortizing it on the basis of the number of kilowatt-hours sold annually. The more energy sold, the lower the capital cost per unit; conversely, the less sold, the higher the capital cost per unit. Some analysts estimate that a capacity factor of 55 percent causes a 36 percent higher capital cost than a capacity factor of 75 percent.[2]

Although there are exceptions, nuclear power plants generally are not impressive from the standpoint of reliability as measured by the capacity factor. For example, according to economist Charles Komanoff, an economist and a student of capacity factors, the unit-year average through 1977 was only 59.8 percent for nuclear power plants. The following table was compiled by Komanoff.[3] (Figures are given in percentages.)

	Year 1976	Through 1976	Year 1977	Through 1977
Unit-year average	57.5	58.5	63.9	59.8
Number of units	48.0	48.0	51.0	51.0
Number of unit-years	48.0	161.0	51.0	212.0

Nuclear-power-plant performance varies according to the type of reactor and the region, but perhaps most significantly it varies according to size. Komanoff states that through 1977, plants of under 800-megawatt capacity have averaged 65 percent capacity factor, but plants larger than that have averaged only 53 percent. That fact is particularly important when one bears in mind that most of the proposed new nuclear power plants not yet operating are in the 1,000-megawatt range. I was struck by Komanoff's findings that larger nuclear power plants were actually less reliable than smaller ones. My whole assumption that they benefited from economies of scale was undercut by this discovery.

An important reason for the poor capacity factors of nuclear power plants is that when there is a malfunction, the danger of massive amounts of radiation in and around the reactor dictates the shutting down of the plant to make necessary repairs. Repair work is slow and difficult, due in part to the need to protect workers from excessive radiation, which necessitates appropriate suiting up and rotation of workers to minimize exposure. For example, in 1974 a cooling pipe at Consolidated Edison's Indian Point plant broke, forcing a repair job that took seven months, 700 men, and $1 million. According to an executive of that company, a comparable repair in a fossil-fuel plant would have taken only 25 workers and two weeks to repair.

Uncertain dependability of nuclear power plants and long repair times cost ratepayers money. As a result of the accident at Three Mile Island, the utilities involved had to purchase power from other sources at the cost of $800,000 per day, and a large number of customers were forced to pay higher utility bills as a result.

Fuel Costs

The rising cost of the fuel needed to run a nuclear power plant relates both to the availability of uranium (found mainly in the western and southwestern United States) and the reliability of the complex system needed to make it usable. Between 1972 and 1979 the cost of "yellow-cake" (the powdery product of uranium mining and milling) rose from about $6 per pound to $42 per pound. Because the supply of uranium will influence its future price, that issue merits some discussion here.

If the demand for a product goes up and the supply of it remains roughly constant, its price should naturally rise. As of August 1979, the generating capacity (or demand) of nuclear power was 52,437 megawatts, according to the U.S. Department of Energy (DOE). In a 1979 letter to the author, that department projected a demand increase of up to 300,000 megawatts by the year 2000. With estimates of 920,000 tons of known uranium reserves and 1,505,000 of probable potential resources, the DOE calculated that the uranium in the United States would be depleted by the middle of the twenty-first century at the latest. As it becomes more scarce, uranium is also becoming more expensive, thus forcing the United States into the position of buying it from other countries. (Indeed, recent Commerce Department figures indicate that U.S. firms paid $43.6 million to the Soviet Union for uranium purchased in 1980.)[4] Although some observers believe the breeder reactor could assure the nation of an indefinite supply of fuel, its development is doubtful because of safety and security reasons, as well as high costs.

Coupled with a shrinking supply of uranium is a complex and expensive system of fuel production that includes the following steps:

1. <u>Mining and milling</u>: The uranium ore must be removed from the ground and then milled, a process in which the ore is crushed and ground wet into a slurry from which the uranium is removed, usually by an acid leach, then purified, filtered, dried, and packaged for transport.

2. <u>Conversion</u>: The uranium must be converted to uranium hexafloride (UF_6), a gas that prepares it for the next step.

3. <u>Enrichment</u>: This stage basically sorts out the uranium 238, which is not fissionable, from the uranium 235, which is. Enrichment commonly occurs by pumping the gaseous UF_6 through a series of porous barriers, each of which discriminates against the isotope U-238.

4. <u>Fuel processing and fabrication</u>: The enriched fuel is transformed back into solid form, shaped into pellets, and placed in stainless steel tubes. Those tubes in turn are placed in fuel rods and inserted into the reactor.

5. <u>Use in the reactor</u>: The fuel rods are used to produce the atomic reaction called nuclear fission, which generates the great heat used to boil water and thus create steam to turn the turbines that create electricity.

6. <u>Disposal of radioactive waste</u>: There is no permanent solution to radioactive-waste disposal, and most of the used, or spent, fuel rods lie in large pools right next to the reactor. In the future, however, something else will have to be done with the waste.

Unexpected delays or interruptions in any of the first four steps could undermine the reliability of the whole commercial nuclear power system. In the judgment of Saunders Miller, author of

The Economics of Nuclear and Coal Power, there are serious questions concerning the adequacy of mining and milling capabilities if demand approaches government forecasts.[5] Similarly, enrichment facilities, owned by the federal government and operated by private companies, may prove inadequate if demand grows apace, according to Miller. Neither the public nor the private sector of the nuclear power industry appears eager to make large investments in enrichment facilities. This may be due in part to the uncertainty of the future of nuclear power itself.

Future Cost of Nuclear Power

Utility companies are already shifting the argument in favor of nuclear power away from cost savings toward the need for a "balanced" approach to energy sources, which would include coal, nuclear, and possibly other power sources. The issue of the appropriate alternatives to nuclear power will be discussed in another chapter. To gain an illuminating perspective on the current and future economics of nuclear power, let us take a look at a test case. The example discussed below was a proposed nuclear facility in western Wisconsin, a good state to look at because its construction and labor costs are about average for the United States.

A CASE STUDY: PROPOSED
TYRONE NUCLEAR POWER PLANT

In 1976 a consortium of providers of electricity in Minnesota and Wisconsin proposed to the Wisconsin Public Service Commission that an 1,100-megawatt nuclear power plant be built in western Wisconsin to serve parts of each state. The dominant member of the so-called Western Wisconsin Utilities (WWU) was Northern States Power Company, which was to have two-thirds ownership of the facility. NSP was the main source of funds and human resources that built the case for the Tyrone power plant. Opposing NSP and WWU was the Wisconsin-based Badger Safe Energy Alliance and the Sierra Club.

Although there were other issues involved in the proceedings with the Public Service Commission (PSC), the final ruling by that body was made very largely on the basis of the economics of nuclear power. The main task of the PSC was to evaluate the cost estimates of those supporting the proposed power plant and those who opposed

In comparing the economics of various alternatives, the costs that occur at various points in time are made comparable to their present worth by the use of this discount rate. The total present worth of each alternative is then directly comparable to the others. Additionally, the total present worth of each plan may be converted to an equivalent annuity (using the same discount rate) over the period of the study. This permits the costs of each alternative to be expressed in terms of annual costs, which are also directly comparable.

Nearly three-fourths of the difference in total capital-cost estimates made by WWU and the opponents of the Tyrone power plant was due to differences in escalation-rate assumptions. The coalition of utilities assumed an annual escalation rate of 7 percent, and the economist for the opponents of Tyrone assumed one of 13 percent. Despite an escalation rate in the 1970s of about 20 percent, the utilities felt that inflation would be reduced greatly in the future, and they anticipated no significantly costly design changes. The anti-Tyrone economist assumed current escalation rates at the time to be "worst case" figures for cost increases and a 7 percent figure to be a "best case" estimate, leading him to what he considered a more realistic 13 percent figure.

The utilities and their opponents also differed considerably on capacity-factor assumptions. The WWU coalition pointed to better-than-average capacity-factor experience with NSP's Minnesota nuclear power plants and decided on the figure of 70 to 75 percent as its assumed capacity factor for the life of the plant. Once again, the economist for the interveners in this case was not that optimistic. He assumed a 60 percent capacity factor, very close to Komanoff's national average annual figure through 1977 of 59.8 percent (a figure that came out after the Tyrone proceedings had ended). The anti-Tyrone economist also said that the proposed Westinghouse reactor had an even worse capacity-factor record than the 60 percent he was assuming. In their rebuttal, the utilities said that a 60 percent average lifetime capacity factor for the Tyrone pressurized water reactor and for coal alternatives struck them as pessimistic, though not inconsistent with historical experience.

Finally, the two sides also differed significantly on their assumptions about the appropriate assumed discount rate. Once again, the coalition of utilities was estimating the lower figure, 10 percent, compared with a range of from 12 to 20 percent submitted by the Sierra Club. Even one percentage point of difference in the estimate would eventually be translated into cost differences of millions of dollars over the life of the plant.

it. While much of the debate centered on nuclear vers[cut]
a look simply at the two sides's conflicting estimates o[cut]
nuclear power will suffice for our present purposes. '[cut]
tions behind the cost estimates are the source of the di[cut]
clusions the two sides reached. While there were num[cut]
ponents of the total cost of the proposed project, both [cut]
edged that differences in estimates of three critical fa[cut]
enced the final verdict on the true cost of the power pl[cut]
three components were escalation rate, capacity facto[cut]
count rate. The escalation rate includes the impact o[cut]
new design requirements on the total construction cos[cut]
facility. Capacity factor measures the power plant's [cut]
measuring real power generation against its potential [cut]
rate is a tool of economic analysis that enables one t[cut]
penditures at different points in time on an equivalent [cut]
allows cost forecasters to establish estimates that a[cut]
ate than would otherwise be the case. According to][cut]
<u>rate</u> can be most simply defined as the rate of compo[cut]
that, when applied to an amount at one point in time,'[cut]
amount that is financially equivalent at another point [cut]
lustrate, assume that you keep all of your money in [cut]
that earns 6 percent interest and that you have purch[cut]
lator from a store that offers the following two paym[cut]
$100 today or $106 one year from today. If you choo[cut]
tion, you would withdraw $100 from the bank and pa[cut]
you choose the second option, you would keep the $1[cut]
and in one year withdraw $106 to pay the bill. How[cut]
year, the $100 has earned $6 interest. Therefore, [cut]
rate were 6 percent, you would conclude that these [cut]
ternatives are financially equivalent.

The discount rate NSP uses is the composite [cut]
The composite cost of capital is the cost of funds s[cut]
preferred stock, and common stock investors. Th[cut]
ently determined as follows:

	Cost (in percent)	Capitalization Ratio
Debt	8.65	.47
Preferred	8.04	.11
Common	15.00	.42
Total		

SURPRISING ECONOMICS / 35

To summarize, the Western Wisconsin Utilities and NSP assumed a 70 or 75 percent capacity factor, a 7 percent capital escalation rate, and a 10 percent discount rate. Based on those and other assumptions, they argued that nuclear power would be less expensive than a coal-powered alternative. The Wisconsin Public Service Commission did not agree with many of those assumptions nor with the conclusion to which those assumptions led.

> The Commission finds that this cost comparison analysis is inadequate to show economic benefit to the system of owning the Tyrone nuclear plant. It contained a number of assumptions as to cost of capital, capacity factor, capital cost . . . and discount rate which the commission considers to be unjustified. . . .
> The capacity factor assumed for Tyrone in these studies is 75 percent. In the absence of working experience with any pressurized water reactor of a size comparable to Tyrone which operates at a capacity factor as high as this, the commission finds applicants' assumptions of a 75 percent capacity factor unreasonably optimistic. . . . Applicants used a 10 percent discount rate for its comparison which is lower than its composite cost of capital; higher discount rate would be appropriate. The commission finds applicants' cost comparisons inadequate to disturb the commission's finding in the advance plan order that coal-fired generation is likely to be less costly than nuclear generation.[6]

The rejected Tyrone nuclear power plant illustrates the new economic realities of nuclear power. While other public service commissions in other states may approve future applications for construction of new nuclear power plants, it is interesting to note that in all of 1978 there were only two new orders for commercial nuclear reactors and none in 1979. In the past five years there have been more cancellations than new orders. To be sure, some of the reduction in the growth of nuclear power is due to the substantial drop in the growth of demand for electricity nationally, but much of the decline in reactor sales must be a result of the rising costs of nuclear power as well.

The Tyrone case also shows the complexity of the job of calculating future costs of electrical power and the need to assess the various assumptions made by the two sides. A cynic might say that it all depends on which experts one wants to believe, and that the evidence presented leaves one with no clear sense of the economic competitiveness of nuclear power. If one is pronuclear, one will make optimistic assumptions about the escalation rate, capacity factor, and discount rate; if one is antinuclear, the opposite will be true. If the assumptions made by the conflicting sides at Tyrone are indicative of such differences throughout the nation, however, the weight of the evidence appears clearly to favor the antinuclear perspective, at least with regard to cost. An escalation rate of 13 percent, particularly in view of Three Mile Island, seems much more realistic than the 7 percent presented by the utilities. A capacity factor of 60 percent seems much more in line with experience than one of 70 to 75 percent. The Wisconsin PSC even found the Sierra Club's assumptions about the discount rate a shade closer to their view of a reasonable estimate.

Despite the soaring costs of nuclear power, however, government and utility company leaders stress the need for nuclear power. Although somewhat greater caution is being exercised by those in a position to order new nuclear reactors, some utility companies persist in advocating nuclear power, still citing low cost as the reason. The rising costs of this technology have drastically changed the economic realities from the standpoint of future ratepayers. What is more, there are hidden costs that have been, are, and will be borne by the taxpayer. When those hidden costs are fairly accounted for, it becomes evident that nuclear power is far from a bargain.

HIDDEN COSTS OF NUCLEAR POWER

The monthly bill paid by the ratepayer is but the tip of the iceberg of nuclear power's true cost. The citizen/consumer as taxpayer has been supporting this technology for nearly four decades. Some estimates of total federal subsidy to nuclear power are as high as $37 billion. Without that support, nuclear power could never have been developed. Those who oppose massive government spending and subsidies or other "unnatural" governmental intervention into the workings of a free economy should be appalled by the way this technology was developed and promoted by the federal government. The government's loan guarantee of the Chrysler Corporation is a minor and fleeting financial commitment compared with its massive and continuous subsidy of the nuclear power industry.

Past, present, and future hidden costs of nuclear power primarily involve various kinds of tax subsidies and support. However, some of the hidden costs, such as those related to radioactive-waste disposal, may involve the citizen/consumer both as taxpayer and ratepayer, as will potential costs related to power-plant accidents. An understanding of all the costs of nuclear power are necessary in making a rational decision about its continued development, for it is the individual taxpayer or ratepayer who foots the bill.

Compared with the energy it produces, nuclear power has been the most heavily subsidized major energy source in the United States. According to a U.S. Department of Energy report, nuclear power accounted for 4.38 percent of all domestically generated energy.[7] However, a 1978 study by Batelle Pacific Northwest Laboratories indicated that over the past 30 years about 13 percent of total federal energy incentives went toward stimulating the development and use of nuclear power.[8] On the other hand, coal, currently the predominant alternative to nuclear power for central-station generation of electricity, enjoyed only 4.5 percent of total federal energy incentives, despite the fact that it accounts for a far higher percentage of national energy output than does nuclear power. Put frankly, the nation has not gotten much of a bang for its nuclear-energy buck.

Federal expenditures for commercial nuclear power have been considerable, and the Batelle study documents past support by taxpayers of nuclear energy. Among the incentives (which we will consider hidden costs for our purposes) cataloged by Batelle were expenditures for research and development of nuclear power, government subsidies to the uranium industry, uranium-enrichment services, regulatory costs, and the insurance of nuclear power by means of the Price-Anderson Act.[9]

THE BATELLE STUDY

Research and Development

Of the $17.1 billion of incentives between 1950 and 1978 documented by Batelle, $15.3 billion was spent on research and development (R&D) of commercial nuclear power. Table 2.1 from the Batelle report shows how dramatically expenditures for commercial nuclear power have increased since 1950, compared with expenditures for all atomic purposes, including military spending. The table includes funds for R&D of uranium enrichment and radioactive-waste disposal, the latter constituting an increasingly large proportion of total expenditures since the mid-1970s. There are some problems separating military from commercial R&D, as the following excerpt from the Batelle study shows:

TABLE 2.1

Federal Funding of Commercial Nuclear Power, 1950–77
(in millions of dollars)

Fiscal Year	Total Fiscal Year Cost,[a] Funds Appropriated (net)	Portion for Nuclear Power[b]	Percent Total	Portion for Civilian Nuclear 1976 Dollars
1950	702.9	7.2	1.0	17.0
1951	2,032.1	14.8	0.7	32.4
1952	1,605.7	20.3	1.3	43.5
1953	4,126.5[c]	23.8	0.6	50.7
1954	1,042.5	38.5	3.7	81.5
1955	1,209.9	57.1	4.7	121.4
1956	834.2[d]	82.6	9.9	173.0
1957	1,898.7	135.8	7.2	274.7
1958	2,334.0	165.5	7.1	325.9
1959	2,635.0	199.5	7.6	389.6
1960	2,649.6	260.8	9.8	501.2
1961	2,666.8	262.2	9.8	499.0
1962	2,547.3	251.2	9.9	472.8
1963	3,134.8	253.2	8.1	470.7
1964	2,742.7	235.0	8.6	431.2
1965	2,624.5	239.2	9.0	431.5
1966	2,433.0	230.6	9.5	404.5
1967	2,438.6	235.8	9.7	402.0

Year			
1968	2,497.0	288.9	11.6	472.6
1969	2,550.6	277.5	10.9	431.0
1970	2,493.7	284.5	11.4	417.1
1971	2,494.6	332.1	13.3	466.9
1972	2,551.6	404.3	15.8	550.3
1973	2,646.8	498.6	18.8	638.8
1974	2,724.9	642.3	23.6	741.2
1975	3,362.8	846.1	25.2	895.2
1976	4,063.4	929.1	22.9	929.1
1976 to	1,171.8	309.0	26.4	309.0
1977[c]	5,350.1	1,427.6	26.7	1,427.6

Total in 1976 dollars 12,401.40

[a]1950–59: Atomic Energy Commission, Annual Financial Report, 1959; idem, Annual Financial Report, 1965; idem, Annual Financial Report, 1975; 1976–77 estimate: idem, Authorizing Legislation, Hearing, Joint Committee, 1977, p. 383.

[b]J. N. Longston, Nuclear Energy Branch, Office of the Controller, Energy Research and Development Administration.

[c]Estimate

[d]Includes transfer to operations of $571.0 million appropriated in prior years as plant and equipment funds.

Source: Batelle Pacific Northwest, An Analysis of Federal Incentives Used to Stimulate Energy Production, prepared for the U.S. Department of Energy, December 1, 1978, p. 120.

In the early years of atomic energy the weapons program developed many aspects of the merging commercial nuclear power program. Methods of handling radioactive materials, neutron diffusion codes, critical experiment technology, and other information were largely applicable to the commercial program. The commercial program developed around an alternative fuel form (uranium oxide rather than uranium metal), cladding material, pressure member (vessel rather than tube), moderator (light water rather than graphite or heavy water), and reactor components. Technology from these developments became available to the weapons program. Fuel reprocessing technology, as presently conceived for commercial nuclear power, is based on weapons program-developed processes, but it is not clear at this time that these processes will become commercial. Waste management technology is being developed for both applications.

Out of the military reactor program grew the pressurized water reactor technology. But again, fuel forms differ, reactor components are substantially larger and are of different designs for the commercial market. Compactness and long-life are much more important to military applications. Further, much of the military technology remains classified while most of the commercial technology is reported in the open literature and thus is available for military application.[10]

A common misconception is that nuclear power was developed in the 1950s and that the R&D costs of this technology are largely behind us. The Batelle study points out vividly that these costs have been growing substantially, not diminishing, now that the basic technology has been developed.

Subsidies of the Uranium Industry

According to the Batelle study, the uranium industry has been influenced to a greater extent by government policy than has any other natural-resource industry. In the early years of nuclear power just after World War II, the government wanted to stimulate the development of uranium resources beyond the estimated reserves, so it offered domestic producers long-term contracts with the following incentives: a ten-year guaranteed minimum price for certain high-grade uranium ore, a $10,000 bonus for the discovery and production of high-grade uranium ore, and a guaranteed three-year minimum price for ores from the Colorado Plateau. The incentive worked: 261,000 minable tons of contained U_3O_8 were discovered between 1948 and 1958. But then the government had to deal with problems of excessive stockpiles of uranium, and it discouraged further exploration in the late 1950s. In order to get companies to agree to defer their sales of uranium, the government agreed to assure the companies a minimum price of $8.00 per pound of U_3O_8 if they would prolong the period during which they sold it. It became evident in the early 1960s that, in order to help prop up demand for uranium and keep the industry alive, the United States would need to prolong the program through 1970. After 1971 the uranium-producing industry was able to make it in the commercial market without government support. However, the government did continue to restrict the importation of foreign uranium.

In 1971, however, there remained 50,000 tons of surplus U_3O_8 owned by the Atomic Energy Commission (AEC). The AEC decided to dispose of the stockpile through a system, called the split-tails plan, which provided that customers of the AEC for enriched uranium deliver only 80 percent of the uranium needed to produce the desired amount of enriched uranium, with the other 20 percent to come from AEC stockpiles. In return, the government's customers would pay 25 percent more for enriching services than was actually delivered. A strong case can be made that this practice constituted a subsidy of the uranium industry, since the cost of uranium has gone up at a faster rate than has the cost of enrichment services during the 1970s. The stockpiles were scheduled to be depleted by the end of that decade.

Although governmental actions have had a profound and beneficial effect on the uranium industry, the nature of the subsidies and incentives do not lend themselves to quantification. However, that does not make them less real.

Enrichment Services

Currently, the federal government is the exclusive provider of uranium-enrichment services at three facilities. The government's pricing policy is that it merely recover its costs from its customers. If this phase were done privately, additional costs, such as insurance, taxes, and a provision for return on equity would make it more costly. While there are many potential ways to calculate the cost of subsidizing a function like enrichment, Batelle selected $1.7 billion (1976 dollars) as the incentive on the basis of the $0.5 billion General Accounting Office estimated subsidy of the difference between commercial and government prices, plus the $1.2 billion outlay (not yet recovered) for increasing the enrichment capacity for commercial purposes. Since 1965 the federal government has been supplying utilities with enriched uranium.

The Cost of Regulation

Whether under the auspices of the AEC or the NRC, the cost of regulation has risen dramatically as nuclear power has expanded. The regulatory system encompasses rule making, licensing, and coordination of policy. In constant 1976 dollars, regulatory costs went up from $5.9 million in 1960 to $146.2 million in 1977. In the ten years from 1965 through 1974 the cost of regulatory activities increased over seven times. The running total since 1960 is $1.2 billion, three-fourths of which has come since 1970. What is more, by its very nature, the greater the attempts that are made to ensure the safety of nuclear power, the greater the regulatory costs will be. If the number of nuclear facilities is allowed to grow, so will the cost of regulation grow.

Liability Insurance

Another subsidy that is difficult to quantify is the liability insurance provided for the nuclear power industry by the government through the Price-Anderson Act. In the early years of nuclear power, utilities were skittish about too deep a commitment to that technology because of the potentially high liability they might have in the case of a catastrophe. Private insurers would not touch this kind of insurance because of the technological uncertainties and high potential losses. So, in 1957 Congress passed the Price-Anderson Act, which limited the liability of nuclear-power-related businesses, including chemical processing, fuel fabrication, and transporters of nuclear products, as well as of commercial nuclear power plants.

The law provides government indemnity now in the amount of $65 million per reactor for each nuclear accident above the maximum private liability insurance available, which is $160 million, and it makes coverage available through an industry pool of $335 million.

The hidden subsidy comes in two ways. First, the federal government insures the nuclear power plants beyond the level that the free-market insurance companies are willing to go, up to $560 million dollars. Few, if any, technologies require that kind of government backup. It can be fairly assumed that if nuclear power plants were able to acquire the same amount of additional coverage from private insurers, the cost to the utility would be considerably higher—hence, the hidden subsidy.

Second, and probably more important, the law explicitly limits the total liability of the utility to $560 million in the event of a nuclear accident. In a worst-case scenario that would involve thousands of deaths and many more made ill or diseased by exposure to high levels of radiation, it is likely that the damage to the public would far exceed the limited coverage. The Batelle study explains the limitation of liability in the following statement.

> Without Price-Anderson, the utilities would have to purchase liability insurance. They would also have to estimate a cost for the uncertainty that a potential loss might exceed the liability limits available on the private market. These costs would be passed on to the consumer in higher electricity prices. The price of nuclear power would therefore increase and the utilities would have to decide whether nuclear power could be competitive and profitable in relation to other energy sources.[11]

ADDITIONAL SUBSIDIES AND HIDDEN COSTS

The Batelle report took into account many, but not all, of the hidden costs of nuclear power. A February 1981 article in Not Man Apart reported that the Energy Information Administration (EIA) within the Department of Energy revealed additional subsidies: in a 1980 study an estimated total of 37 billion tax dollars (1979 dollars) had been spent on commercial nuclear power.[12] Among the additional hidden costs were these:

1. The subsidy of reactor sales to other countries under the Agency for International Development, the Export-Import Bank, or other means—estimated cost: $237.4 million;

2. Strong federal assistance in uranium prospecting through the National Uranium Resource Evaluation (NURE) and other subsidies in uranium production—estimated cost: $2.5 billion;

3. Nuclear-waste-management subsidies (part of which is included in Batelle's R&D estimates) including $300 million for the cleanup of mill tailings from abandoned mines and $2.1 billion of spent-fuel-handling costs when that occurs—estimated cost: $6.5 billion.

Two additional subsidies are missing in the EIA report: the subsidies provided by reactor manufacturers early in the development of nuclear power, and tax benefits enjoyed by large nuclear power plants. In the early 1960s, Westinghouse and General Electric entered into turnkey contracts with utilities, such that a reactor would be built for a fixed fee and the manufacturer would assume the risk of all cost overruns or technical failures. Large overruns were apparently assumed by builders of reactors, but it was worth the cost to create a new market.

As for tax subsidies, economist Duane Chapman of Cornell is reported to estimate the total state and federal subsidies to be about $200 million for each new large nuclear plant.[13] Investment tax credits, preferential depreciation schedules, and other tax advantages given such large capital investments account for the bulk of that sum, and a rough calculation of the annual national tax subsidy amounts to about $15 billion.

THE COSTS OF AN ACCIDENT

Some of the potential costs of an accident at a nuclear power plant are difficult to delineate. Unless the severity of the incident is known, there is no way a dollar figure can be attached to it. In a worst-case scenario, the federal government (and thus the taxpayer) could be presented with a bill in the millions or even billions of dollars.

Added to personal liability suits could be decommissioning costs, that is, the cost of permanently closing down and making safe the faulty reactor. While estimates exist for decommissioning costs of reactors that have not been closed owing to an accident, little is known about the costs of premature, accident-caused decommissioning costs. Equally imponderable is the effect on the local economy of a serious nuclear accident. In the case of large urban centers, where massive and prolonged evacuation could be required, that impact is apt to be considerable. Evacuation expenses in the case of Three Mile Island were considerable, but in that case relatively little radiation contaminated the surrounding area, compared with the potential each power plant has. Residents in the Harrisburg area were able to return to their homes quite soon afterward.

Although many of the potential costs that would accompany a serious nuclear-power-plant accident require estimation and speculation, the case of Three Mile Island offers some concrete evidence of the magnitude of risk and potential cost to ratepayers, taxpayers, and even stockholders.

The events at Three Mile Island revealed some of the economic dangers inherent in depending heavily on a technology that requires a protracted shutdown in the event of an accident. The cost of purchased power from other systems has already been noted; it amounted to $800,000 per day and necessitated rate increases for many of the affected customers. A further look at the financial fallout of Three Mile Island is instructive.

General Public Utilities (GPU), owner of the two nuclear units at Three Mile Island, is among the largest utility companies in the United States. Yet one accident pushed it to the edge of a financial precipice. According to GPU's chairman, William Kuhns, the company would have faced insolvency had it not been able to secure a massive loan from a consortium of 43 banks. The utility company received a loan of $409 million to help see it through the crisis, although participating banks had to be promised that they could withdraw from the agreement in case of another serious accident. Like the cost of purchased power, the eventual repayment of the $409 million will, of course, fall on the shoulders of the customers of GPU and its subsidiaries, the Metropolitan Edison and Pennsylvania Electric companies.

Then there was the cost of cleanup at Three Mile Island. Insurance did cover the roughly $1 million in claims resulting from the accident, and insurance would cover up to $300 million to repair damage done to the plant. However, GPU's early estimates indicated that cleaning up the reactor core would cost from $60 to $80 million, and cleaning up everything else would cost over $400 million. Later estimates have now reached $1 billion. Shortly after the accident, a GPU spokesman said they hoped that sister utilities and/or the government would help pay the $160 million not covered by insurance; the rationale for this was that the cleanup operation would yield valuable information for all utilities using nuclear power. A realistic appraisal of the probable source of funds, however, points again toward the ratepayer and/or the taxpayer. GPU also filed a federal-court suit seeking more than $500 million in damages from the plant's designer and supplier, Babcock and Wilcox Company.

One indication of how financially disastrous just one accident was is the change of bond rating given the owners and operators of Three Mile Island. Moody's Investors Service, exactly one year after the accident, reduced the rating of the bonds issued by the utilities that own the crippled power plant, leaving them the lowest-rated utility bonds in the nation. Moody's said that it considered investment in such bonds to be "speculative."

Stockholders of electrical utilities heavily dependent on nuclear power were also hurt by the accident at Three Mile Island. However, a few factors tended to moderate Wall Street's reaction to Three Mile Island.

Electrical utilities had not been doing well as investments for the preceding 15 years, so the stock had a shorter distance to fall.

One old investment maxim is, "Buy on the bad news." Confidence within the investment community that management knows what it is doing helps provide a company with a kind of "safety net" in the event of a one-time calamity.

Utilities are powerful institutional customers for brokerage houses, and even implied criticism of that utility's top management by second-guessing its commitment to a given technology could be embarrassing or worse.

Even with those moderating influences, the stocks of heavily nuclear electrical utilities have demonstrably suffered. In Perception for the Professional, a publication of the brokerage house of Piper, Jaffrey, and Hopwood, a study has been made of 13 heavily nuclear electrical utilities and of 13 nonnuclear electrical utilities. Table 2.2 shows that the pre-Three Mile Island performances of nuclear and nonnuclear stocks were virtually identical, but that after the accident there was a clear difference in performance, a difference that has cost owners of heavily nuclear electrical-utility stock dearly.

Of another Three Mile Island, Steve Leuthold wrote the following.

> Even if it only approaches the seriousness of Three Mile Island, let alone worse, it could bring about the shut-down of existing facilities. This could leave 100 incredibly expensive monuments, many with multi-billion dollar price tags, standing idle. The revenues generated from those plants would stop. But the debt incurred to build these plants would remain. And the interest on the debt would

TABLE 2.2

Wall Street Nuclear Fallout, Performance Comparisons

	12/29/78 to 4/30/80	Pre-Three Mile Island, 12/29/78 to 3/23/79	Post-Three Mile Island, 3/23/79 to 4/30/80
NYSE utilities	-1.2	+3.6	-4.6
13 "no nuke" electrics	-6.0	+2.6	-8.4
13 "nuke" electrics	-13.9	+2.5	-15.9
Excluding General Public Utilities	-7.8	+2.5	-10.0

Source: Steve Leuthold, "Wall Street Nuclear Fallout," Perception for the Professional, written for Piper, Jaffray, and Hopwood, May 1980, p. 4.

continue to run. Seventy-two nuclear power plants now operate and another thirty-two are in the late stages of construction. Most NUKE facilities could be put in dire straits, not just the single company experiencing the accident.[14]

Before Three Mile Island, such an analysis would have been considered hysterical. Now it rings true.

CONCLUSIONS AND COMMENTS

As I studied the economics of nuclear power, several new insights became particularly impressive.

1. The rate of increase both for capital and fuel costs was surprisingly high, far in excess of the rate of inflation. Much of the

capital-cost escalation was due to safety-related requirements, meaning that the trend will undoubtedly persist.

 2. The reliability of nuclear power plants is low. Never having heard of capacity factors before starting my research, I was particularly struck by the length of time nuclear power plants can stand idle for one reason or another.

 3. The fact that nuclear power occurs in an economic environment of regulated monopoly rather than classic free-market competition seems crucial. The result is that there has been little deterrent on the part of utility companies to overbuild and inadequate incentive to keep costs down. The problem is exacerbated by such agreements as the automatic adjustment clause for fuel, which, like many of the other rising costs mentioned earlier, are just passed along to the ratepayer.

 4. The future-cost estimates of nuclear power made by the utility companies in the Tyrone proceedings seemed unreasonably optimistic. To assume an escalation rate of 7 percent for capital costs when the 1970s had experienced an average annual increase of nearly 20 percent seemed far-fetched. Similarly, to assume capacity-factor performance of 10 to 15 percent above average appeared unfounded.

 5. The list of government subsidies, direct or indirect, was striking. I was particularly impressed by the rate of increase of costs of research and development of nuclear power; I had thought that the trend would be in the opposite direction. I was also surprised that, according to the Batelle study, coal had enjoyed only about half the federal subsidies, or incentives, that nuclear power had received.

 6. Some of the subsidies, such as the Price-Anderson insurance coverage given the private nuclear power industry by the federal government, are not quantifiable but could be very large if there were a serious accident. Once again, government involvement has distorted the true cost of the "product" of nuclear power—and potentially to the detriment of the taxpayer.

 7. As a technology, nuclear power is much more vulnerable financially than I had thought, an impression sharpened by the experience of Three Mile Island. The statements by the head of General Public Utilities about possible insolvency were really surprising. Similarly, the devastation of the bond rating of Three Mile Island's owners was revealing. If just one accident can do that to large utility companies, what will be the effect of another Three Mile Island? The fact that heavily nuclear electrical-utility stock had dropped

significantly and for a sustained period after the accident in Harrisburg may be the most eloquent statement of nuclear power's financial fragility.

I was able to present much of the information on the economics of nuclear power during the 1980 session of the Minnesota State Legislature as I continued to advocate my bill on nuclear waste. I had thought that the changed economic position of this technology would make some impression on Republicans, a mistaken belief. The same people who spoke with the greatest fervor on behalf of the free-enterprise system on the floor of the House were the most steadfast defenders of nuclear power, a puzzling fact. The advent of nuclear power bears almost no relationship to the free-enterprise system, but that does not move conservative Republicans to look at it askance.

Issues related to the safety of nuclear power seem to be the most potent in the popular mind and in the mind of many legislators. In order to shore up my Democratic support and keep the bill's chances alive, I felt I had to probe the safety issue. I still tried to remain objective in looking at the facts, and I was suspicious of scare rhetoric on either side of this controversy. The facts I was to encounter, however, would prove to be unsettling.

NOTES

1. John Emschwiller and Robert L. Simison, "Continuing Cloud: Midwest Nuclear Plant, in the Works 13 Years, Keeps Facing Delay," Wall Street Journal, March 4, 1980, p. 1.
2. Saunders Miller, The Economics of Nuclear and Coal Power (New York: Praeger, 1976), p. 68.
3. Charles Komanoff, Nuclear Plant Performance Update 2 (New York: Komanoff Energy Associates, 1979), p. 7.
4. "Soviet Uranium Sold to U.S. for Nuclear Use," Minneapolis Tribune, August 18, 1981.
5. Miller, The Economics, p. 24.
6. Wisconsin, Public Service Commission, "Findings of Fact, Conclusions of Law and Order," March 9, 1979, p. 9.
7. U.S., Department of Energy, Energy Information Administration, Annual Report to Congress, 1979, vol. 2: Data (Washington, D.C.: Government Printing Office, 1979), p. 5.
8. Batelle Pacific Northwest, "An Analysis of Federal Incentives Used to Stimulate Energy Production," December 1, 1978, p. 255.

9. Ibid., pp. 111-51.
10. Ibid., pp. 121-22.
11. Ibid., p. 127.
12. Jim Harding, "The Nuclear Blowdown," Not Man Apart, February 1981, p. 16.
13. Ibid.
14. Steve Leuthold, "Wall Street Nuclear Fallout—More to Come," Perception for the Professional, written for Piper, Jaffray, and Hopwood, May 1979, p. 9.

REFERENCES

Articles

Bupp, Irvin C. "Nuclear Realities." New York Times, May 29, 1979, p. A-48.

Emschwiller, John, and Robert L. Simison. "Continuing Cloud: Midwest Nuclear Plant, in the Works 13 Years, Keeps Facing Delay." Wall Street Journal, March 4, 1980, p. 1.

Harding, Jim. "The Nuclear Blowdown." Not Man Apart, February 1981, p. 16.

Martin, Douglas. "Three Mile Island: Financial Fallout." New York Times, January 13, 1981.

Salisbury, David. "Nationalize Nukes? It Could Happen Here." Christian Science Monitor, May 24, 1979, p. A-43.

Shipman, William D. "When a Nuclear Plant Shuts Down, Consumers Pay." New York Times, May 14, 1979, p. A-43.

Smith, Don S. "Nuclear Power's Effects on Electric Rate-Making." Public Utilities Fortnightly, February 2, 1978, pp. 16-22.

"Unpaid Costs of Electricity." Science News, February 17, 1979, p. 105.

Reports and Miscellaneous Material

Batelle Pacific Northwest. An Analysis of Federal Incentives Used to Stimulate Energy Production. Prepared for the U.S. Department of Energy, December 1978.

Bertschi, Rudolph L. Advance Plan to Wisconsin Public Service Commission. Docket no. 05-EP-1, March 22, 1978.

Ferguson, Robert L. Letter to author, August 13, 1979.

Kellenyi, John. "General Public Utilities Dividend Cut: Observations and Investment Opinion." Research Bulletin of Donaldson, Lufkin, and Jenrette, May 8, 1979. Mimeographed.

Knecht, Ronald. Testimony on Power Generating Economics and Planning to Wisconsin Public Service Commission. Docket no. CA-5447, December 28, 1978.

Komanoff, Charles. Nuclear Plant Performance Update 2. New York: Komanoff Energy Associates, 1979.

Leuthold, Steve. "Wall Street Nuclear Fallout." Perception for the Professional, written for Piper, Jaffray, and Hopwood, May 1980. Mimeographed.

———. "Wall Street Nuclear Fallout—More to Come." Perception for the Professional, written for Piper, Jaffray, and Hopwood, May 1979, p. 9.

North, D. Warner. Exhibit 164 to Wisconsin Public Service Commission. Docket no. CA-5447, January 15, 1979.

Silverstein, Evan J. "Electric Utilities: The Nuclear Risks." Memorandum of L. F. Rothschild, Unterberg, Towbin, May 18, 1979. Mimeographed.

U.S., Department of Energy, Energy Information Administration. Annual Report to Congress, 1979, vol. 2: Data. Washington, D.C.: Government Printing Office, 1979, p. 5.

U.S., General Accounting Office, Comptroller General. Nuclear Power Costs and Subsidies. Washington, D.C., June 13, 1979.

Warwick, J. B. "Electric Utilities: Utilities Anonymous Revisited." Memorandum of Morgan Stanley, May 14, 1979. Mimeographed.

———. "Electric Utility Industry—Random Thoughts on the Long Term Effects of Three Mile Island." Research note of Morgan Stanley, June 18, 1979. Mimeographed.

Wisconsin, Public Service Commission. Findings of Fact, Conclusions of Law and Order. March 9, 1979.

Worner, Gerald. On Reserves and Reliability. Report to Wisconsin Public Service Commission, November 15, 1978.

Worner, Gerald, Edward Cazalet, and D. Warner North. <u>Supplemental Testimony on Economic Studies and Reserves and Reliability</u>. Report to Wisconsin Public Service Commission, January 15, 1979.

Speeches

Miller, Saunders. Speech presented at University of Louisville, June 8, 1979, Louisville, Ky.

Perl, Lewis J. "Estimated Costs of Coal and Nuclear Generation." Speech presented to New York Society of Security Analysts, December 12, 1978, New York.

U.S., House of Representatives, Committee on Interior and Insular Affairs, Subcommittee on Energy and Environment. <u>Comparative Costs of Coal and Nuclear Electricity Generation</u>. Testimony by W. W. Brandforn, July 10, 1979, Washington, D.C.

_____. <u>The Economics of Nuclear Power</u>. Testimony by Charles Komanoff, July 12, 1979, Washington, D.C.

Books

Ford Foundation. <u>Nuclear Power Issues and Choices</u>. Cambridge, Mass.: Ballinger, 1977.

Miller, Saunders. <u>The Economics of Nuclear and Coal Power</u>. New York: Praeger, 1976.

3

LOW-LEVEL RADIATION: HIDDEN DANGER

When we toured the Monticello nuclear power plant in February 1979, I believe each member of the House Energy Committee was aware of its invisible presence. Although we were given small meters to measure it as we made our way through the facility, we were not required to put on any clothing to protect ourselves from it, a fact that surprised me. However, a large device was used to measure our exposure to it as we left the reactor area, ensuring that none of us had become accidentally overexposed. Its presence was felt as acutely even in its apparent absence as any of the sights and sounds we actually did encounter during our informal investigation of the power plant. It was the life stuff of the nuclear genie and we were closer to more of it then than we had ever been before.

Radiation has always seemed exotic to me. This subatomic force is odorless and colorless; low levels of exposure to it cannot be felt; and its true effects on living organisms remain shrouded in mysterious uncertainty. The presence of radiation in the production of nuclear power is what gives this technology the aura of danger it possesses. In the case of the radiation that occurs within a reactor, there is no doubt as to its potentially lethal power if not controlled and contained properly. The safety of low-level radiation, on the other hand, is like a mirage: the more one learns, the farther away a definitive, provable conclusion on its true danger becomes.

In order to learn about low-level radiation, I had to review dry and technical reports, some of which will be referred to in this chapter. I must confess that I could ascertain little consensus on this controversy. I did come to appreciate the fact that the nuclear-fuel production process is fraught with much more danger than I had ever expected. I had believed that the most important risk involved in atomic energy was a possible mishap at a reactor of the kind that occurred at Three Mile Island. The disturbing information I was to come across made me more mindful of the real risks run daily by uranium miners and others in the nuclear-fuel cycle. Much of the literature on the risk of low-level radiation is anecdotal and not purely scientific from the standpoint of establishing cause-and-effect proof of a certain kind of radiation causing cancer. However, the evidence I came across in my reading, whether in the form of scientific studies or mere anecdotes, should not be ignored in assessing the true risks of nuclear power.

A theme I noticed first in researching low-level radiation in the nuclear-fuel cycle and strikingly saw later in looking at reactor safety was that of inadequate management control and safety precaution to a degree that surprised me as a layman. Past practices of allowing uranium miners to work with virtually no safety standards related to radiation exposure, letting people build homes out of highly radioactive material, and the existence of large quantities of material unaccounted for at uranium-enrichment facilities all suggested a system that has inadequately minimized the risks involved in producing usable uranium. While there may well be many competent individuals working within various parts of the nuclear power industry, the system as a whole has been far less error-free than I had assumed. In the case of workers exposed to low-level radiation, however, the significance of those failures is difficult to demonstrate, for the precise danger of such radiation remains a puzzle.

LOW-LEVEL RADIATION: HOW SAFE?

The primary reason nuclear power is dangerous is the presence of the radiation that is associated with it from start to finish. The danger of radioactive substances lies in their effect on living cells, the distortion of which can lead to cancer, birth defects, and mutations. Radiation is of two kinds: electromagnetic (such as X rays) or particulate (emitting alpha or beta particles). The dosage of radiation is measured in rems or rads, and the discussion of allowable limits of exposure to radioactive substances is generally in terms of millirems or millirads (measuring one-thousandth of a rem or rad). The <u>rem</u> refers to the amount of radiation required to produce a particular amount of biological damage in tissue. The rad specifies the amount of energy absorbed by tissue. Radioactive materials emit ionizing radiation, so called because it separates electrons from atoms, thus producing ions, or atoms with an electrical charge. (For a further discussion of ionizing radiation, see Appendix C.)

There is no controversy with regard to large doses of radioactivity. A dose of 500 rem will cause death to about half of the population group exposed to it; 250 rem will cause a large fraction of the exposed population to die and many others to become sick. No one questions the extreme danger of the highly radioactive fuel used in a reactor; such fuel sends out tens of thousands of rems. With low-level radiation, however, the picture is less clear.

In 1963 the Atomic Energy Commission (AEC) decided to reassess the existing standards of acceptable radiation exposure in response to growing misgivings in the scientific community. Two respected scientists, John W. Gofman and Arthur R. Tamplin, agreed to investigate the accepted standards. Gofman, a medical physicist, was an associate director of the Biomedical Research Division at the AEC's Lawrence Radiation Laboratory at Livermore, California. Among other accomplishments, Gofman had distinguished himself as a codiscoverer of four radioactive isotopes. Tamplin had a doctorate in biophysics and was a group leader in Gofman's division. Their study was initiated by the AEC; thus no antinuclear bias can be imputed to Gofman and Tamplin before they began their research.

The messengers brought bad news. On October 29, 1969, Gofman and Tamplin presented their findings to the Institute for Electrical and Electronic Engineers in San Francisco. At that time the Federal Radiation Council had guidelines for acceptable radiation exposure: no individual in the general population should receive more than .5 rads per year, and the average annual dose should not exceed .17 rads (standards that persist today). Gofman and Tamplin told the audience, "If the average exposure of the U.S. population were to reach the allowable 0.17 rads per year average, there would, in time, be an excess of 32,000 cases of fatal cancer plus leukemia per year, and this would occur year after year."[1] Their findings were met with incredulity in some quarters. As Gofman and Tamplin stated, "Many of the people, in nuclear electricity work, simply expressed their disbelief that a Federal Agency would ever set a guideline that could be associated with such an enormous hazard. They all had been under the illusion that the hazard at the guideline radiation level must be zero, or at least so very low as to be negligible. One after another officials of the nuclear electricity industry expressed their opinion that surely something must be wrong with our estimates, although not one of them could muster an iota of evidence as to what it could be."[2]

The response to Gofman and Tamplin's findings on the part of nuclear power supporters went beyond incredulity. Some simply denied the validity of their findings, without pointing out any errors or shortcomings of the two scientists' methodology. Others, like Theos Thompson, a member of the U.S. Atomic Energy Commission, would defend the standards with statements such as this, made to participants at the AEC's briefing May 21, 1970, on nuclear development in Columbia, South Carolina: "Obviously this is a very small amount of radiation compared with the levels which mankind has been receiving through all of the ages. To date, in spite of many careful studies, no one has been able to detect any effect from these low levels of radiation and it is unlikely that studies of literally millions of cases would show any such effects."

The rhetoric used to attack Gofman and Tamplin's findings has never been matched by a corresponding scientific case disproving their results. In that sense, the controversy surrounding their findings is different from other studies dealing with occupational exposure to low-level radiation.

Probably the most prominent controversy on the health effects of low-level, occupational radiation centers on Thomas Mancuso and his associates Alice Stewart and George Kneale (see Chapter 7). Mancuso began studying the causes of death for workers at the U.S. government's nuclear facilities in Hanford, Washington, in 1964. After over 10 years of study, he determined that 6 percent of the deaths of Hanford workers were caused by radiation, a finding the U.S. Department of Energy (DOE) strongly disagreed with. In fact, in 1977 the federal government withdrew its financial support of Mancuso's research. Critics of Mancuso's study believe he used defective statistical methods, citing as reasons the small population that he studied (35,000 workers), the fact that measured leukemias were not of the expected magnitude, and that the increased cancers could have been unrelated to radiation. Mancuso and his associates digested the criticism and reanalyzed their studies, only to reach their original conclusions. Respected scientists disagree on this controversy and on others involving radiation exposure in the United States, from Rocky Flats, Colorado, to Portsmouth, New Hampshire. All such debates have been inconclusive for reasons that will soon be evident. The differences might best be summed up by Arthur Upton, former director of the National Cancer Institute:

"Fragmentary and incomplete data do not by sheer numbers make a case."[3] On the other hand, Mancuso states, "When a series of independent studies point in the same direction, the evidence should not be ignored."[4]

Yet hard and fast conclusions on the health effects of low-level radiation are elusive. Nothing illustrates that fact better than a 1979 report by the National Academy of Sciences (NAS). This document of over 600 pages demonstrates more what is not known than what is agreed upon regarding low-level radiation and its effects on people. To be sure, there is some consensus on a few points.[5]

1. Cancer is the main sickness that arises from excessive radiation exposure, although there is great variety in the susceptibility of various organs to radiation-induced cell damage.

2. Excess cancer from radiation exposure can vary according to sex.

3. Age is an important factor for radiation-induced cancer, and fetuses are particularly vulnerable to excessive exposure.

4. Genetic effects of excessive radiation exposure have not been proved for human beings, and estimation of those risks must depend on experimentation on laboratory animals. How transferrable to man are the findings related to other animals remains an open question. The report does not indicate that radiation exposure has figured prominently in observed genetic abnormalities to date.

5. The biggest source of man-made radiation is X rays. Safely operating nuclear power plants account for relatively little radiation to the general public.[6]

6. Strontium 90 and Cesium 137 are the two most potentially dangerous by-products of nuclear fission from the standpoint of an accident exposing them to the general population.[7]

The areas of uncertainty in the NAS report are striking.

1. What is the shape of the dose-response curve at low levels of exposure? The report theorizes that for some kinds of radiation a linear dose-response relationship may understate the risk and for other kinds of radiation it may overstate them.

2. What mathematical assumptions should be made in making estimates and what procedures are appropriate? What is the dimension of the uncertainties involved?

3. What effect does the dose rate have on the radiation effect?

4. Do X rays or gamma rays of 100 millirads per year present a danger to those exposed to them?

These questions are illustrative of the nature and degree of unanswered questions that came out of the NAS report, a document that represented the most recent and comprehensive attempt to answer the question of how dangerous low-level radiation truly is. Indeed, there was considerable controversy within the panel of scientists that developed the final report, another indicator of how little has been irrefutably proved on this subject.

Bearing in mind that numerical estimates depend on the assumptions made, and that there is nowhere near unanimity on what those assumptions should be, here are some examples of the range of estimates, depending on the dose-response model and the kind of risk-projection model used for exposure to X rays or gamma rays.

For a one-time exposure to 10 rads of X-ray or gamma radiation, deaths beyond those expected in a given population group range from 0.47 percent to 3.1 percent, according to the report. For a lifetime exposure to one rad per year of the same kind of radiation, the range is from 2.8 percent to 18.2 percent excess deaths.[8] The organs of women are twice as likely to fall prey to cancer as those of men from 11 to 30 years after radiation exposure.[9] Based on the findings of this report, it is clear that even low levels of exposure to radiation are not healthful and should be avoided as much as possible.

Still, the report did not come up with proof that a given low level of radiation is of a specified magnitude of danger. A further look will show just how the cards are stacked against any such proof.

In June 1979 the Interagency Task Force on the Health Effects of Ionizing Radiation issued its findings and commented on the problems of assessing the effects of low-level radiation. Those problems are considerable.[10]

1. It is impossible to determine whether any given cancer was caused by radiation or something else. Cancer has a long latency period, and merely looking at cancerous tissue does not reveal the source.

2. Confounding variables, such as a chemical carcinogen like benzene, can distort a reading of cause and effect in a dosage-response analysis.

3. Information on doses to the entire body or to specific organs is not always available, so again there are problems analyzing dose and response.

4. It is often difficult to determine the dose received by an individual, and as a result it may be difficult to tell if an observed

response is greater than would be expected if the actual dosage was low.

5. Record keeping is a problem. Doses of radiation may never have been recorded, and subsequent health-outcomes information may be unavailable (perhaps due to privacy laws), inaccessible, or incomplete.

6. The available sample size may be too small. According to the report, "In the case of low doses of radiation, conclusive demonstration of effects might require studying a population of a million or more." None of the populations studied to date reaches this magnitude, and most are substantially smaller.

7. In order to compare the effects on an exposed group, there must be a control group with generally similar characteristics, but lacking the exposure. In the case of a large sample size of the kind needed to assess the effects of low-level radiation, this may be impossible to assemble.

8. Key demographic and exposure factors must be known. The kind of radiation, the dose level and dose rate, as well as the age, sex, and health status of the people affected must be known.

It is evident that to get all of the necessary variables properly lined up to undertake a conclusive experiment on low-level radiation requires an extraordinary effort, many people, and a great deal of time. An attempt at such a study will be a large-scale follow-up of workers in seven shipyards where nuclear work has been done, which is to be undertaken by researchers at Johns Hopkins University over five years and at a cost of $10 million. Mancuso is already critical of the proposed study because most nuclear shipyards are relatively new, and the latent period for cancer is typically 15 to 20 years or more; he believes such a study requires 20 years. It appears doubtful in any case, for the reasons cited by the Interagency Task Force, that an irrefutable finding will be reached on this subject.

There remains an open question, Is it correct to assume that low-level radiation should be considered safe until absolutely proved dangerous? Or should it be considered dangerous until proved safe? A look at different parts of the nuclear-fuel cycle and the people exposed to higher-than-normal levels of radiation is warranted.

RADIATION DANGERS IN THE NUCLEAR-FUEL CYCLE

Each of the steps involved in producing and disposing of nuclear fuel poses some level of risk because uranium waste products emit unusually high amounts of radiation. To illustrate that point, let us take a look at uranium mining, milling, enrichment, fuel fabrication, and transportation.

Uranium Mining

Like coal miners, uranium miners in this country's 280 mines are susceptible to respiratory disease and prone to accidents, but unlike other kinds of miners, they face an additional danger—radiation. Much of the danger comes from the radon gas present in uranium mines. As the radon decays, it produces a series of isotopes called radon daughters, which attach to dust particles and can become lodged in miners' lungs. Radon daughters are alpha-emitting particles that commonly attack lung tissue, leading eventually to cancer. The concentration of radon is measured in terms of "working levels," a measure of the amount of radioactive material to which the miners are exposed. To measure exposure over time, scientists use a unit called Working Level Month (WLM); there is no general conversion of WLMs to rems or rads, because WLMs are based on the location of the mine, its airflow patterns, and other factors. It has only been since the late 1950s that there have been miner-exposure radiation standards, and since 1967 the limit for underground miners has been 4 WLMs per year, a standard erratically enforced until the federal government, through the Mine Safety and Health Administration, began to do so in 1972.

The absence of decent standards has taken its toll on uranium miners. In a joint monograph done for the Department of Health, Education and Welfare (HEW), three researchers found a statistically significant excess of respiratory cancer in miners who had experienced lifetime, cumulative exposures of from 120 to 359 WLMs.[11] Interestingly, the researchers, F. E. Lundin, J. K. Wagoner, and V. E. Archer, found that the risk of respiratory cancer per unit of exposure was greater in low cumulative-radiation groups than in higher ones, a finding that would suggest that a linear or proportional relationship between dose and resulting cancer may not, in some cases, be a conservative enough basis for setting effective radiation standards.

Another study found that uranium miners in Canada, Czechoslovakia, and the United States were one-and-one-half to two times more likely to get cancer than was the general population. Researchers Archer and Wagoner, joined by J. Dean Gilliam and Lynn A. James in a study on uranium miners of the Colorado Plateau,[12] reported a significant excess of respiratory cancer among both whites and Indians and went on to point out that standards for airborne noxious agents should consider the amount of physical labor involved, the size of the individual workers, and the height above sea level.

Because these studies include miners who have worked in unregulated conditions, it is difficult to assess the adequacy of current standards; one way to determine that would be to have a comprehensive health study of miners who have worked only since the institution of the current government standards. Moreover, even with the current standards, there are problems of radon-daughter control and effective monitoring of exposure. Max B. Slade, a health specialist with the U.S. Mine Safety and Health Administration, has pointed to the following problems with radon-daughter control: reopened mines with shafts parallel to old ones are subject to sudden influxes of radon gas; additional radon gas can come from mill tailings used for backfill material to aid in ground control; and most mines are not laid out for radon control, leading to problems in sealing off worked out areas. Monitoring and enforcement difficulties include the practice by some companies of averaging WLM exposure over all the miners so that some will have received more than they should have; variations in sampling frequency; failure of smaller mining concerns to own sampling equipment; and human error, such as when miners turn off ventilating fans for their own comfort or inadvertently block air-circulation passages.[13]

Other evidence of the dangers of uranium mining is anecdotal. For example, according to Tom Barry of the Navajo Times, 25 Navajo uranium miners in the Red Rock Chapter of the Navajo Reservation have died of lung cancer after working the Kerr-McGee mines in Red Rock and Cove, Arizona.[14] When he published the article on August 24, 1978, 20 more were reportedly dying of the same disease. In his article, Barry quoted LuVerne Hussen, Public Health Service Director at Shiprock, New Mexico, "It was a get rich scheme that took advantage of Navajo miners who didn't know what radioactivity was or anything about its hazards. Inside the mines were like radium chambers, giving off unmeasured and unregulated amounts of radiation. The problem was that back in the 50's nobody was riding herd on the companies. The uranium mine operators got what they could as quick as they could out of those mines. They sent anybody who was old enough to hold a shovel and handle a wheelbarrow into the mines to cart the stuff out."[15]

The adequacy of current standards remains an open question. A task group of the National Council of Radiation Protection and

Management has been studying the issue and is apt to encounter problems of methodology like those already mentioned. If there is uncertainty regarding the adequacy of government standards for uranium miners today, there can be no question of the callousness shown in the past by the federal government and the uranium-mining companies vis-à-vis uranium miners. Not only was there a failure to push for rigorous safety standards but miners were allowed to remain so ignorant of the hazards of their work that in some cases they built their homes from the highly radioactive rocks of the uranium mines, according to the "ABC News Closeup: Power—The Uranium Factor," televised in April 1980. For example, in Grand Junction, Colorado, at the old Climax Uranium Company, uranium tailings were taken away by individuals to be used in building construction. Before this practice was stopped in 1966, 800 structures were contaminated. The same program showed evidence that one large uranium-mining company, United Nuclear Corporation, encouraged its employees to misrepresent the amount of radiation to which they had been exposed. Indians in the western states have often been victims in the past and will continue to be in the future because about half of the nation's uranium reserves are on Indian lands.

Nor is the impact of uranium mining limited to those who mine it: the environmental impact is considerable. Mining modifies the normal groundwater-flow cycle below the water table, and the lower water table can expose mineralized rocks to a new environment, leading to the oxidation and dissolving of radiochemical and toxic materials. Air quality is affected, too. Radon is emitted from shafts into the surrounding air, and some of the daughter radioactive isotopes appear to enter the food chain. Other problems involve waste rock, land use, noise, and conventional air and water pollution.

Uranium Milling

Milling involves crushing the rock and dissolving it through a solvent-extraction process producing "yellowcake," a semirefined uranium compound. The most serious environmental-health problem comes from the tailings that accompany the milling process. However, there are also problems on the work site. The major hazards to uranium millworkers are airborne silica, uranium, vanadium, acids, and alkalis. Another study done by Archer, Wagoner, and Lundin on smoking and uranium mining also found an excessive amount of malignant disease of the lymphatic and hematopoietic tissue among millworkers. Four cases of lymphatic cancer were discovered, when only one case is to be expected statistically in a normal population group.[16]

In terms of public health, mill tailings provide a potentially considerable danger. According to a draft of the Work Group on Radiation Exposure of the HEW Interagency Task Force on Ionizing Radiation, there are 27 million tons of tailings at 24 inactive milling sites and another 113 million tons at active uranium-milling locations.[17] If the radioactivity is not completely contained, it can contaminate the environment through radon exposure to the lungs by inhalation; through whole body irradiation directly from the pile; through the deposit of radionuclides in the body of a person who ingests contaminated food or water; and possibly through exposure to radon daughters and radium if mill tailings are used in construction or in landfill, as has occurred in Colorado. A tailing pile gives off radon 222 and its daughter products, along with radium 226 and thorium 230, thus providing both a gamma and alpha radiation hazard for up to 80,000 years, according to some sources.

According to an April 1979 study on uranium milling by the Nuclear Regulatory Commission (NRC), if most past milling practices were followed into the future, the odds that a person living close to a mill for 20 years would die prematurely of cancer would be 600 in a million, an estimate that leaves large margin for error.[18] The estimated risk is about 40 percent higher than the risks that background radiation provides. Moreover, the NRC concluded that by the year 2000 people living in a region where milling and mining are carried on together face a risk about double that posed by milling alone.

The dangers are not merely hypothetical. For example, in Grants, New Mexico, there was a major breach in the tailings impoundment on February 5, 1977, which caused some 50,000 tons of radioactive slimes and solids and several million gallons of contaminated water to spill over an area of 60 acres of land.

Similarly, on July 16, 1979, the dam that contained much of the mill waste from United Nuclear's Church-Rock facility broke, causing 97,000 tons of the material to spread out over the landscape, resulting in radiation of 10 times the normal level in the neighboring sheep. The federal government can allow state governments to regulate this aspect of nuclear power, and in the case of New Mexico, it does. However, New Mexico lacked the staff to detect that United Nuclear failed to follow the originally approved design of the dam; that the dam had been cracking for a long time, a fact known to United Nuclear; and that United Nuclear built the dam higher, even though it was already cracked. Such an incident raises the question of whether the NRC should directly assume the responsibility of regulating the operation of all uranium mills to help ensure adequate management of such facilities.

Uranium Enrichment

The process of enriching uranium for use as fuel is carried on in three government-owned, industry-run facilities that can generate enough enriched uranium for 90 reactors. These facilities use much water and enormous quantities of electricity (according to some estimates, as much as 3 percent of all U.S.-produced electricity), and exposure to radiation in such facilities can be high. Chemical accidents, fires, or explosions can compound that seriously. In terms of public exposure to radiation, the Environmental Protection Agency estimates that .007 percent of the material processed is released into the biosphere, and if that is correct, about 7,000 tons of such material escapes from a facility that produces about 100 megatons of solid wastes per year.[19] Uranium and daughter products pass into the air and water. The Environmental Protection Agency (EPA) is not allowed to assess the costs and efficiencies of systems to remove gaseous and liquid wastes for security reasons, so it is difficult to know how much more this part of the nuclear-fuel cycle could be cleaned up.

There is evidence that the three enrichment facilities may have have contaminated the surrounding countryside. A petition to the NRC in 1978 quoted Charles Thornton, former AEC director of Nuclear Materials Safeguards, as saying the following:

> The aggregate MUF (material unaccounted for) from the three diffusion plants alone is expressible in tons. No one knows where it is. None of it may have been stolen, but the balances don't close. . . . The AEC can say officially that quantities of MUF are not dangerous. That is not so. Tons have been lost. They can say they have impregnable barriers, sensitive modern instruments. Not that impregnable, not that sensitive. They can say "The numbers are not good, but we don't know how to do better." If you admit that this industry is not controllable, then you shut down. You wait until it is controllable, and then start up.[20]

The same petition, filed by Jeannine Honicker, cites an epidemiological study that found higher-than-expected cancer rates in Tennessee's Roane and Anderson counties, where the Oak Ridge Gaseous Diffusion Plant is located. The study showed that nonwhite females in Anderson County died of lung cancer and leukemia between 1950 and 1969 at four times the rate expected from the vital statistics of the state. According to the study, residents of the two counties have a higher age-adjusted, race-factored, and sex-factored mortality in 22 of 30 categories compared with Tennesseans as a whole.

Fuel Fabrication

Fuel fabrication, the most costly part of making atomic fuel, consists of converting uranium hexaflouride to uranium dioxide (UO_2), solidifying the UO_2 into small ceramic pellets, and loading them into metal fuel rods. Workers in the eight fuel-fabrication facilities in this country encounter low-level gamma radiation and suspended radioactive particles in the air. According to Tucker's study, "The NRC estimates a critical organ dose commitment of 9,950 millirem to the lungs of fuel fabrication workers, and a whole body dose of about 260 millirem, which is double natural background radiation. The DOE has reported that fuel 'processing' accounts for the largest individual exposure to workers, averaging 540 millirem."[21]

The people in the surrounding countryside of at least one fuel-fabrication plant appear to have been affected by radioactive releases from it. The Tucker study asserted that the Getty Nuclear Fuel Services plant near Jonesboro, Tennessee, received attention when the <u>Atlanta Journal</u> reported that cancer deaths in the affected county doubled in the 20 years since the plant opened in 1958. A sharp increase was reported in 1973 after a 15-year latency period had passed. Drinking water in the area showed a positive reading of radioactivity, and 250 to 500 pounds of enriched uranium was reportedly released into the Nolichucky River in 1977 alone.[22]

Transportation

From the mines to the fabrication plants to the reactors, uranium travels thousands of miles in this country, regulated by the Department of Transportation. According to the HEW Interagency

Task Force, approximately 2.5 million packages of radioactive materials of all kinds were shipped in 1974. The results of occupational exposure are not well documented, but there are examples of accidents that should be of concern to the local residents affected. According to the Tucker study, Critical Mass Journal reported in December 1977 an accident that occurred in southeastern Colorado, spilling 10,000 pounds of yellowcake over a 5,000-foot area. It was caused by several horses bolting in front of a truck carrying 40,000 pounds of yellowcake, the semirefined milled uranium. Poorly trained and poorly equipped personnel showed up 12 hours later to begin cleaning up the yellowcake, some of which was a foot deep. The public was told of the wreck a week later. According to the same article, nearly 250 other truck accidents involving radioactive material had occurred since 1971.[23]

A news report in another paper, the Sunday Oklahoman, reported an accident six days before Three Mile Island. This time a semitrailer jackknifed, overturned, and about 20 of the 55 gallon drums full of yellowcake broke. The uranium was spread far and wide by high winds through the vicinity of Wichita, Kansas.[24]

Even if the quality of containers is improved and additional safety steps are taken, it is likely that transportation accidents will occur. But, as with the affected citizens of eastern Colorado or Kansas, in many cases it will be nearly impossible to prove that a specific accident was the cause of cancer many years later.

CONCLUSIONS AND COMMENTS

    1. I was struck by how difficult it is to prove or disprove the danger of low-level radiation. The methodological barriers to establishing any ironclad proof seemed to be such that it will be a long time before there is a consensus on this issue.

    2. In the meantime, it is generally assumed that unless low-level radiation can be proved to be dangerous, it should be considered safe. If that assessment is warranted, so much the better; but if not, some of the possible long-term genetic effects of exposure to such radiation are troubling, not to mention the short-term effects (see pp. 167-69).

    3. The people who work on various stages of making uranium into useful nuclear power fuel are subject to risks, the true magnitude of which is not known, in large part because of the uncertainties of the effects of low-level radiation. I was particularly struck by the dangers uranium miners are exposed to. Standards have

risen considerably in recent years, but it seems to me that uranium miners remain the most vulnerable workers in the whole nuclear-fuel cycle.

4. The story of John Gofman and his odyssey from government-supported researcher to vehement nuclear power critic is of great importance. If nuclear power grows as much as the industry hopes it will, I suspect Gofman's view that there is no threshold of safe exposure to low-level radiation will become appreciated belatedly.

Although the safety record of the nuclear-fuel cycle is spotty at best, and in the past uranium miners have been shamefully exposed to high levels of radiation, there appeared to me to be little potential for a broad scale catastrophe in the process of making uranium usable for nuclear fission. That does not bring back to life the victims of poor management and inadequate caution shown in the past by this part of the nuclear power industry. The subtlety of the dangers of low-level radiation, however, may explain the somewhat haphazard quality of preventing accidents and preserving the public health and safety, as it regards uranium production. I had assumed the story would be very different when I began to look at reactor safety, the most publicized aspect of the nuclear power safety issue.

NOTES

1. John W. Gofman and Arthur R. Tamplin, Poisoned Power: The Case against Nuclear Power Plants (Emmaus, Pa.: Rodale Press, 1971), p. 69.
2. Ibid., p. 98.
3. Jean L. Marx, "Low-Level Radiation: Just How Bad Is It?" Science, April 13, 1979, p. 164.
4. Ibid.
5. National Academy of Sciences, Committee on the Biological Effects of Ionizing Radiations, Division of Medical Sciences, Assembly of Life Sciences, National Research Council, The Effects on Populations of Exposure to Low Levels of Ionizing Radiation, Washington, D.C., 1980, pp. 3-9.
6. Ibid., p. 51.
7. Ibid., p. 61.
8. Ibid., p. 193.
9. Ibid., p. 250.

10. U.S., Department of Health, Education and Welfare, Report of the Interagency Task Force on the Health Effects of Ionizing Radiation, Washington, D.C., June 1979, pp. 29, 30.

11. Kitty Tucker, Uranium and the Nuclear Cycle (Washington, D.C.: Health and Learning Project, 1979), p. 20.

12. U.S., Department of Health, Education and Welfare, National Institute for Occupational Safety and Health, Center for Disease Control, Epidemiological Studies of Lung Cancer among Uranium Miners of the Colorado Plateau, from the Conference/Workshop on Lung Cancer Epidemiology and Industrial Applications of Sputum Cytology, Golden, Colo., November 14-16, 1978, pp. 180-81.

13. Ibid., pp. 452-61.

14. Tom Barry, "The Navajo Lung Cancer Widows," Navajo Times, August 24, 1978, p. B13.

15. Ibid., p. B14.

16. Tucker, Uranium, p. 21.

17. U.S., Department of Health, Education and Welfare, Report of the Interagency Task Force, p. 27.

18. U.S., Nuclear Regulatory Commission, Draft Generic Environmental Impact Statement on Uranium Milling, NUREG-0511, Washington, D.C., April 1979, pp. 4-5.

19. U.S., Environmental Protection Agency, Office of Radiation Programs, Environmental Analysis of the Uranium Fuel Cycle: Part 1—Fuel Supply, EPA-520/9-73-003-13, 26, Washington, D.C., 1973, p. 103.

20. Tucker, Uranium, p. 41.

21. Ibid., p. 50.

22. Ibid., p. 51.

23. Ibid., p. 56.

24. "Spilled Uranium Ore Cleaned from Road," Sunday Oklahoman, March 25, 1979, p. 13.

REFERENCES

Articles

Archer, V. E., J. K. Wagoner, and F. E. Lundin. "Uranium Mining and Cigarette Smoking Effects on Man." Journal of Occupational Medicine, vol. 204 (1973).

Barry, Tom. "The Navajo Lung Cancer Widows." Navajo Times, August 24, 1978, pp. B13-14.

Marx, Jean L. "Low-Level Radiation: Just How Bad Is It?" Science, April 13, 1979, pp. 160-61, 204.

Morgan, Karl Z. "Cancer and Low-Level Radiation." Bulletin of Atomic Scientists, September 1978, pp. 30-40.

"Spilled Uranium Ore Cleaned from Road." Sunday Oklahoman, March 25, 1979, p. 13.

Reports and Miscellaneous Material

American Broadcasting Corporation. "ABC News Closeup: Power —The Uranium Factor." April 23, 1980.

Lederberg, Joshua. "Affidavit before the Public Service Board of Vermont." Docket no. 3445, September 18, 1970.

National Academy of Sciences, Committee on the Biological Effects of Ionizing Radiations, Division of Medical Sciences, Assembly of Life Sciences, National Research Council. The Effects on Populations of Exposure to Low Levels of Ionizing Radiation. Washington, D.C., 1980.

Tucker, Kitty. Uranium and the Nuclear Cycle. Washington, D.C.: Health and Learning Project, 1979.

U.S., Department of Health, Education and Welfare. Report of the Interagency Task Force on the Health Effects of Ionizing Radiation. Washington, D.C., June 1979, p. 27.

U. S., Department of Health, Education and Welfare, National Institute for Occupational Safety and Health, Center for Disease Control. Epidemiological Studies of Lung Cancer among Uranium Miners of the Colorado Plateau. From the Conference/Workshop on Lung Cancer Epidemiology and Industrial Applications of Sputum Cytology, Golden, Colo., November 14-16, 1978, pp. 180-81.

U. S., Environmental Protection Agency, Office of Radiation Programs. Environmental Analysis of the Uranium Fuel Cycle: Part 1—Fuel Supply. EPA-520/9-73-003-13 26, Washington, D. C., 1973, p. 103.

U. S., Nuclear Regulatory Commission. Draft Generic Environment Impact Statement on Uranium Milling. NUREG-0511, Washington, D. C., April 1979, pp. 4-5.

Books

Beckmann, Petr. The Health Hazards of NOT Going Nuclear. Boulder, Colo.: Golem Press, 1976.

Gofman, John W. "Irrevy": An Irreverent, Illustrated View of Nuclear Power. San Francisco: Committee for Nuclear Responsibility, 1979.

Gofman, John W., and Arthur R. Tamplin. Poisoned Power: The Case against Nuclear Power Plants. Emmaus, Pa.: Rodale Press, 1971.

4

REACTOR ACCIDENTS: CLEAR AND PRESENT DANGER

I must confess that I approached the subject of reactor safety with trepidation. I am not an engineer and knew that I would be out of my depth if I tried to become an expert on the technology of nuclear reactors. My limitations became painfully clear to me when I tried to understand and explain exactly what had happened at Three Mile Island. A friend of mine, Winthrop Rockwell, a lawyer in Minneapolis who served on the staff of the President's Commission on the Accident at Three Mile Island, was kind enough to review my manuscript dealing with that subject and recommended that I deal with the technical issues in a general way and avoid getting bogged down in detail. It was good advice and I have followed it. I do believe that an educated judgment on the safety of nuclear reactors can be made, however, even without being a nuclear engineer. Let me explain.

A look at the record of nuclear accidents in the past few years, as well as the system of accident prevention that is supposed to avert episodes like Three Mile Island, allows a person to compile enough information to form an educated impression of the safety of nuclear power. Engineers may be able to look at a piece of equipment or a structure and determine whether or not it is defective or likely to become so. As a layman, I am not able to do that and have not tried. But I have accumulated information from many sources, including the Nuclear Regulatory Commission (NRC), which have helped me piece together enough of the puzzle to see what the picture is. Indeed, not being an engineer may be an advantage in this effort; all I can really understand is information that describes results and the system at work that attempts to ensure reactor safety.

It will be useful, I hope, to list some of my preconceptions on this subject, since I think they may well represent those of many other people as well.

1. No previous accident had even approached the seriousness of the one at Three Mile Island.
2. Nuclear power plants were closely regulated by federal authorities who possessed and used stiff enforcement tools to ensure maximum possible public safety.
3. Operators of nuclear power plants were universally well trained by an instructional system that had high standards and was certified by the federal regulators.
4. The only danger in an accident at a nuclear power plant came in the malfunction or damage of a major component; a small part such as a valve could not cause a disastrous chain of events.
5. Once an accident occurred, the sophisticated monitoring equipment in a reactor would detect it immediately, and plant operators would know what to do about the problem once it was detected.
6. There was good communication within the industry such that a malfunction at one plant would be known to operators of other similar plants, and appropriate caution would be exercised.

In general, nuclear power was perceived to be a technology of the highest possible order of sophistication, run mainly by the private sector to ensure the highest possible level of management efficiency. It was a tight system that worked. If man could invent it, man could control it.

AN OVERVIEW OF NUCLEAR REACTORS

Generating electricity through nuclear fission has been called an exotic means of boiling water, and that is indeed the primary function of a nuclear reactor. The heat released by fission turns water into steam, which then drives a turbine to rotate a generator that creates electricity. Once the heat has been generated, there is fundamentally no difference in how electricity is created between a plant that is coal fired and one that is nuclear powered.

A typical nuclear reactor is bullet shaped, about 40 feet high and 15 feet in diameter, and surrounded by steel walls at least eight

REACTOR ACCIDENTS / 75

FIGURE 4.1

Boiling Water Reactor (BWR)

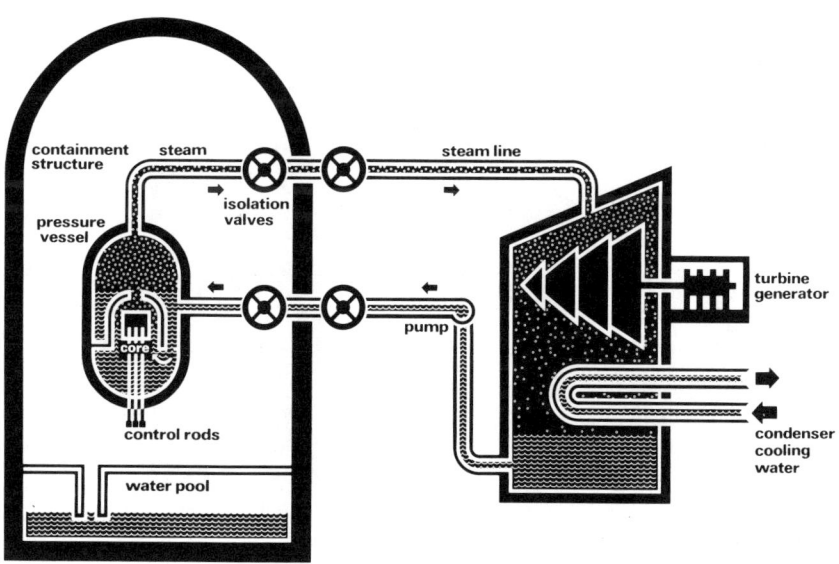

Source: Reprinted, by permission, from Atomic Industrial Forum.

inches thick. It contains roughly 100 tons of uranium fuel assemblies suspended in racks within the chamber. The fuel is in the form of half-inch cylindrical pellets, which are packed in 12-foot-long tubes made of zirconium-alloy metal with spaces between them to allow water to circulate adequately to control the atomic reaction.

In the United States the two kinds of commercial nuclear reactors are boiling water reactors (BWRs) and pressurized water reactors (PWRs) (see Figures 4.1 and 4.2).

One difference between the two kinds of reactors lies in the pressure exerted on the water in the reactor vessel: a BWR exerts about 1,000 pounds per square inch and a PWR about 2,500 pounds per square inch. The greater the pressure, the higher the temperature required to make it boil.

76 / NUCLEAR ENERGY IN THE UNITED STATES

FIGURE 4.2

Pressurized Water Reactor (PWR)

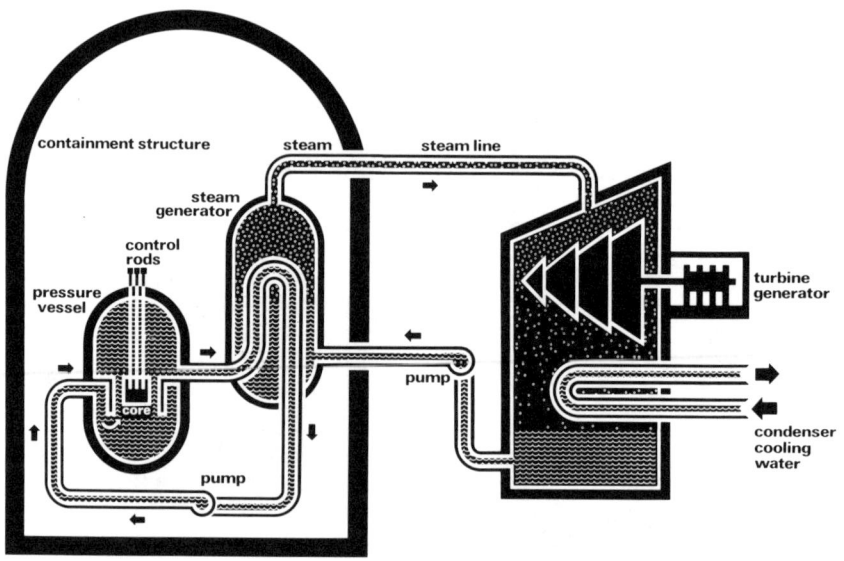

Source: Reprinted, by permission, from Atomic Industrial Forum.

In a boiling water reactor, steam passes directly from the reactor core through a steam line to the turbine. In a pressurized water reactor, the water is piped to a steam generator, where it flows through many thin tubes that in turn heat a separate stream of water, creating the steam that moves through the steam line to the turbine. In both kinds of reactors, after the steam turns the turbines to generate electricity, it is restored to a liquid state by means of a condenser and is returned either to the reactor vessel (in a BWR) or to the steam generator (in a PWR).

In either kind of reactor it is essential to control the nuclear fission. Rods of boron or cadmium are inserted into the reactor and serve to absorb the neutrons that fission causes to move throughout the area of atomic reaction. The raising and lowering of these control rods regulates the rate of fission, and in the case of a reactor accident the control rods are supposed to be lowered automatically. Water is also used to remove heat from the fissioning fuel, which can exceed 4,000° F at its center. Water is needed even after the control rods have been lowered because some of the fission by-products (such as strontium) continue to decay radioactively and can generate enough heat to melt the core if the cooling water is lost.

Because the cooling of the core is so important, each reactor has a backup cooling system. If the normal cooling system fails, the backup system is designed to flood the core and pressure vessel with cool water of such a chemical content as to be able to absorb the speeding neutrons. If that system fails, the Emergency Core Cooling System would flood the reactor with more water.

Failure of the backup, or redundant, cooling systems is what could lead to a worst-case nuclear accident. A loss of water would bring controlled fission to a halt, but the heat caused by the fission by-products accumulated in the uncooled reactor would cause the fuel rods to heat up enormously at a rate of as much as 400° F every 10 seconds. In less than a minute, the fuel rods, surrounding metals, and everything else in the reactor would collect in a glob of molten steel, uranium, and radioactive poisons, and within an hour that material would melt its way through the reactor vessel. This is what is called a meltdown, and it has the catastrophic potential to contaminate the air and water for many miles around the plant. The extent of contamination is a function of the water table and prevailing winds in the affected area.

But the massive release of highly radioactive material to the area surrounding a nuclear reactor need not occur only by means of a meltdown. Serious accidents at nuclear power plants with never-anticipated causes have occurred in ways that have threatened the surrounding populations.

REACTOR ACCIDENTS: THEY HAPPEN

The Nuclear Regulatory Commission does not like to use the word accidents in describing things that go wrong at nuclear reactors; they describe them as "incidents," which are minor, and "abnormal occurrences." An abnormal occurrence is defined by the NRC as "an unscheduled incident or event which the Commission determines significant from the standpoint of public health and safety."

TABLE 4.1

List of Abnormal Occurrences at Nuclear Power Plants
(since January 1975)

Description	Date
1. Steam-generator-tube failure	February 26, 1975
2. Fire in electrical-cable trays	March 22, 1975
3. Loss of main-coolant-pump seals	May 1-2, 1975
4. Improper control-rod withdrawals	May 3, 1975
5. Pipe cracks in boiling water reactors	September 1974 to January 1975
6. Fuel-channel-box wear	April 17, 1975
7. Feedwater-flow instability—water hammer	Up to May 1975
8. Deficiencies in containment	January 1976
9. 8-rem occupational whole-body exposure	March 19, 1976
10. 10-rem occupational whole-body exposure	April 5, 1976
11. Failure of undervoltage-trip logic and consequent loss of safeguard AC power	July 1976
12. Nuclear-core power distribution anomaly	July 8, 1976
13. Steam-generator-tube integrity	September 1976

14.	Improper control-rod withdrawals and unplanned reactor criticality	November 12, 1976
15.	Feedwater nozzle cracking in boiling water reactors	1974 to March 15, 1977
16.	Breach of physical security system	April 18-19, 1977
17.	Fuel-rod failures at nuclear power reactor	May 15, 1977
18.	Management and procedural-control deficiencies	July 1977
19.	Generic-design deficiency	August 10, 1977
20.	Environmental qualifications of safety-related electrical equipment inside containment	November 1977
21.	Insulation failures in containment electrical penetrations	September 1977 to November 1977
22.	Fuel assembly control-rod guide-tube integrity	December 13, 1977
23.	Overexposure of two radiation-protection technicians	April 6, 1978
24.	Degraded primary-coolant boundary in a boiling water reactor	June 17, 1978
25.	Loss of containment integrity	July 1978 to September 1978
26.	Degraded engineered-safety features	September 16, 1978
27.	Deficiencies in piping design	October 1978
28.	Nuclear accident at Three Mile Island	March 28, 1979

Source: U.S., Nuclear Regulatory Commission, "List of Abnormal Occurrences at Nuclear Power Plants," Washington, D.C.: Government Printing Office, July 1979.

Thousands of incidents have been reported to the NRC, and as Table 4.1 indicates, 28 abnormal occurrences were reported between February 2, 1975, and March 28, 1979.

A brief look at three accidents, two of which are on the list (numbers 2 and 28), is illustrative of the dangers and risks inherent in this complex technology characterized by the presence of massive amounts of radiation. The first accident occurred in the Detroit, Michigan, area; the second in the area near Decatur, Alabama; and the third at Three Mile Island near Harrisburg, Pennsylvania.

Detroit, Michigan, 1966

On the afternoon of October 5, the Enrico Fermi breeder reactor experienced partial melting of its fuel when the sodium-cooling system jammed. A breeder reactor is different from the kind generally used in the United States because its fuel is made up of plutonium and uranium 238. When the fission of the plutonium occurs, neutrons enter the nonfissionable U-238, actually creating, or "breeding" new plutonium. Because of the large quantities of plutonium present in such a reactor, extreme caution must be exercised in the operation of the plant. Ironically, at the Fermi plant it was a metal triangle—not in the blueprints—which was installed to provide an additional safety device, that was ultimately discovered blocking one of the coolant nozzles.

The emergency-scramming device, which automatically turns off the reactor, closed the plant down and averted a runaway reaction, but for months it was feared that the fuel would melt further. There was even talk of a possible evacuation of 1.5 million residents of Detroit. It took months to determine how the accident had occurred, and it was a year and a half before the metal object that caused it was removed. It was four years before the reactor could operate again, and in 1972 its license was revoked and it was closed permanently.

Decatur, Alabama, 1975

At the Browns Ferry reactor on March 22, two electrical technicians were checking for air leaks in the cable-spreading room, where electrical cables used to control the reactor are routed into the correct buildings. Unfortunately, they used a candle in their attempt to locate the leaks and accidentally lit the flexible polyurethane-foam sealing material, causing flames quickly to spread to the electrical cables themselves. From there the fire spread to the reactor building, where it burned for seven hours before it was brought under control. The fire crippled several crucial parts of the Emergency Core Cooling System, so that in one of the reactors there was a period during which neither the normal cooling system nor the emergency core cooling system worked, raising the possibility of a meltdown. Fortunately, an auxiliary pump could be used through a makeshift connection to force the needed water into the reactor.

Plant management apparently had not appreciated the dangers of using flexible polyurethane sealing material, even though several small fires had occurred before in similar testing. Furthermore, personnel inspecting the sealing had not been provided adequate written procedures. The fire caused the reactor to be shut down for nearly a year and a half and cost the ratepayers $18 million per month in replacement power.

Harrisburg, Pennsylvania, 1979

The accident at Three Mile Island on March 28 began with a malfunction in the system that feeds water through the reactor, which caused the turbine generator to turn off automatically. This caused pressure in the system to rise, triggering the opening of a valve to release it. When the pressurizer relief valve failed to close appropraitely, which was undetected for over two hours, an opening was created in the primary coolant system, which led to the most dangerous kind of reactor accident—a loss-of-coolant accident. The nuclear fuel rods received inadequate coolant, causing some of them to melt and release a large amount of radiation, which triggered a site emergency and then a general emergency. That night, one of the main-reactor coolant pumps was started, and flow through the reactor was again established.

Both units of Three Mile Island were shut down, at a cost to ratepayers of $800,000 per day for purchased power, in an accident that, had there been proper communication, could well have been avoided. Both the manufacturer of the reactor, Babcock and Wilcox, and the Nuclear Regulatory Commission had been warned of problems with the valve that caused the loss-of-coolant accident.

In September 1977 the same kind of valve on a Babcock and Wilcox reactor stuck open at the Davis-Besse plant, causing a chain of events similar to that at Three Mile Island; fortunately, the operators realized in time that it was open and no serious loss-of-coolant accident occurred. Both the NRC and Babcock and Wilcox investigated the incident, but they provided no information about it to the utilities before the accident at Three Mile Island.

An analysis by Carlyle Michelson, an engineer for Tennessee Valley Authority, one of Babcock and Wilcox's customers, also pointed out the problem with the valve and the potential resulting problems for those operating the reactor. Babcock and Wilcox and the NRC both had the opportunity to examine the so-called "Michelson Report," yet neither followed up with the instructions to those operating similar Babcock and Wilcox reactors to enlighten them about the potential problem with the valve or how to recognize and then handle problems caused by its malfunction.

The three accidents cited here demonstrate three kinds of potential failures in operating nuclear power plants: failure of technology, failure of management, and failure of plant personnel.

Failure of some mechanical parts of a nuclear power plant must be expected over the course of its life. Although one cannot predict when it will happen or which mechanism will be affected, some sort of technical failure is inevitable. At the Fermi reactor, a coolant nozzle became plugged, and at Three Mile Island a valve did not work properly. In most cases, such failures do not lead to a chain of events that pose a danger to plant personnel or the public, and sophisticated monitoring of the reactor can nip a potential calamity in the bud. In a few cases, however, a mechanical failure can go undetected for so long (Three Mile Island) or be of such an unpredictable nature (the Fermi reactor) that it takes months to understand what even happened. The complexity of this technology compounds the problem of appropriately fixing a mechanical breakdown, as does the vast radiation associated with it.

Failure of those managing a nuclear power plant or manufacturing reactors can help cause accidents that would otherwise be avoidable. Despite previous, smaller fires, the management of the Browns Ferry reactor failed to establish clear procedures that could have averted the massive fire that eventually occurred. Similarly, Babcock and Wilcox showed poor management in failing to relay to the appropriate utilities the information received on possible problems with the kind of valve whose malfunction led to the events at Three Mile Island. Once the accident near Harrisburg was under way, the firm operating the facility, Metropolitan Edison Company, failed to follow its own procedure of closing a mechanism called the "block valve" once a leak in the pressurizer relief valve occurred, another example of mismanagement.

The case that the accident near Harrisburg was largely owing to errors in management has been made forcefully by Winthrop Rockwell, former associate chief counsel to the President's Commission on the Accident at Three Mile Island (TMI). Rockwell states: "What has been given less attention, perhaps because we are still so close to the details of the accident, is that the accident reflected a basic failure on the part of the organizations involved to manage complex technology safely. Many of the factors contributing to the accident were not merely one-time aberrations, but were the result of basic weaknesses in the systems and management tools used to assure safety in the operation of TMI."[1]

Of many management tools available to reduce the risk of mechanical failure, two of the most useful would have been <u>problem trending</u>, a means of identifying problems that are repetitive and finding out why they recur, and <u>failure analysis</u>, a way of isolating the impact on the whole system of a specific deviation of an individual part. Neither tool was used, nor was there an effective quality-assurance program, according to Rockwell. Indeed, nearly every management tool was used inadequately. Rockwell stated the following:

> To keep the probabilities of failure at an acceptable level of risk (whatever that may be)—whether in nuclear plants, space flight, production of toxic chemicals, aviation or any other safety-critical industry—requires industrial management that is unrelenting and meticulous in its use of the tools that can lower probabilities of failure: design review, testing, quality control, problem trending, failure analysis, redundancy, maintenance, training, instruction procedures, and many more. Almost all of those human systems were found wanting at TMI, and yet without them it is meaningless to talk about "safe" or "well designed" equipment, since no complex machine can operate in isolation from such "intensive care" support systems.[2]

The standards Rockwell establishes are correctly high. Adequate use of all the management tools he refers to would be impressive even if they were applied to only one reactor. Passing the test for 72 reactors, however, would appear most difficult.

Failures on the part of plant personnel can also pose a serious problem. In hindsight, analyzers generally agreed that, had the operators of Three Mile Island adequately used information available to them as the accident unfolded, they could have responded much more quickly. The danger signals were sufficient to trigger an appropriate response. However, when 200 alarms go off in the first two minutes of an accident, as happened at Three Mile Island, and when the control room is not set up in a way to facilitate the identification of the problem, operator error is not hard to understand. Yet a valve failure similar to the one experienced at Three Mile Island had been detected and responded to much more quickly when it occurred in a different reactor a year and a half before the accident at Harrisburg.

It is clear that more needs to be done to prevent accidents of the kind that have been described here. Regardless of future growth or curtailment of nuclear power, there are existing, functioning nuclear power plants that are not going to be closed soon.

PREVENTING AND MANAGING ACCIDENTS

Operator Training

Three Mile Island showed how a small malfunction can lead to a near catastrophe. With inadequate training, power-plant operators can—and have—made bad situations worse, and recent accidents lead to the question of how well operators are trained to deal with abnormal occurrences. Information on the adequacy of power-plant operator training is fragmentary at best, but the findings of one newspaper reporter, Jim Dawson of the Minneapolis Star, are unsettling.[3] Dawson's story of May 9, 1979, indicates that the quality of training is generally uneven and certainly inadequate for managing a reactor crisis.

Northern States Power, which is headquartered in Minneapolis, sends potential plant operators of their Prairie Island facility to the Westinghouse Nuclear Training Center in Zion, Illinois, for part of their training. According to veteran instructors, there are many inexperienced staff people at Zion, using poorly written or inaccurate material, and teaching in a demonstration facility that is in violation of almost every minimum standard for instruction established by Westinghouse. One instructor there had never worked in a commercial nuclear power plant, although he was instructing others on how to do so. Instructors are supposed to be licensed on a neighboring, working commercial reactor within a year of joining the instructional staff; but many of the instructors at Zion do not have that license. The agreement on teaching standards between the NRC and Westinghouse requires a licensed instructor to monitor unlicensed instructors, although that requirement is commonly ignored. One instructor estimated that only 10 percent of the instructional materials were of high quality, and efforts to improve the quality have been rebuffed by cost-conscious managers. Some of the materials are actually inaccurate, according to one instructor.

The NRC does not regulate the operations of the nuclear training centers and only judges the quality of the program on the basis of the "end product," although trainees come from diversified experiences before entering the Zion program. The NRC determines the students' competence when it gives its licensing examination, a test that trainees must take after additional training from the utility for which they are to work. Many companies prefer to do much of the training themselves, thus ensuring a variety of standards and an unevenness of levels of competence among plant operators throughout the country.

The flaws of the training program become magnified when one looks at how to teach operators to manage accidents. Instructors with limited personal experience under normal conditions cannot be expected to know how to advise trainees on what to do under abnormal conditions. Furthermore, there are accidents that involve developments that are unprecedented and lead to a chain of events that even the most seasoned operator does not know how to handle. Surprisingly, nuclear-power-plant trainees are not required to pass a minimum competency test on a simulator as is required of commercial airline pilots before they are qualified to fly. This, too, weakens their ability to respond appropriately to an accident.

The President's Commission on Three Mile Island was most critical of operator training. As the report of October 1979 put it, "As the evidence accumulated it became clear that the fundamental problems are people-related problems, and not equipment problems."[4] The commission called the training of TMI operators "greatly deficient" and said that inadequate attention was paid to dealing with the kind of potentially serious accident that confronted them on March 28, 1979. Fragmented training of personnel was blamed for some of the problem; the oversight on the part of the NRC, the commission said, "may actually aggravate the problem."[5] The report pointed out that the NRC has limited staff to supervise operator licensing, and many of those who do the licensing have no actual experience in power plants themselves.

To improve operator training, the commission recommended that the NRC upgrade its operator and supervisor licensing function, including accreditation of training institutions. Training programs should include emergency and simulator training and instruction in basic principles of reactor science and safety. Training institutions should be subject to periodic review. Operators should be examined and licensed by the restructured NRC.

Information Gathering and Assessment

In order to avert major accidents, it is desirable to have complete and accurate reports on small malfunctions, or "unscheduled events," as the NRC calls them. Nuclear reactors are extraordinarily complex, and, as we have seen, the correct functioning of each component can be crucial. The NRC requires reports of unscheduled events from all of its licensees. In the case of utilities using Babcock and Wilcox reactors, there was a total of 1,672 events in the period 1969 to 1979, and Babcock and Wilcox is only one of several reactor manufacturers. The NRC has no lack of raw information; the key issue is what is done with it to help promote the safety of power plants. A report in January 1979 by the U.S. General Accounting Office (GAO) pinpoints some problems in NRC oversight of commercial nuclear power.[6]

According to the GAO report, the NRC gathers and assimilates the information well enough. At NRC regional headquarters, each licensee event report is examined to assess the correctness of response to the event, to analyze the event's relationship to other systems and components in the power plant, and to determine whether the event needs to be reported to the Congress as an abnormal occurrence. Staffs of the three NRC headquarters offices examine each report to assess its safety importance to the power plant and its possible applicability to other power plants. But as the GAO report says of the NRC, it has not clearly defined these assessment elements for identifying potential safety-related problems from licensee event reports. Rather, it has left to each of the three headquarters offices and five regional offices the discretion of deciding on the scope and frequency of analyses necessary to identify new safety problems as early as possible. The offices have further delegated this decision making to individual staff members. Furthermore, neither NRC as a whole, nor its respective staff offices, has established decision documentation and disposition procedures. Thus, only when a new safety problem is identified can NRC be assured that the report or series of reports has received an adequate assessment.

The need for a more systematic assessment process with clearly defined procedures is illustrated in this report. A 1973

NRC inspection team found a high incidence of inoperative shock absorbers, which are attached to pipes to prevent vibrations from cracking them. The regional NRC officials informed officials at headquarters of the problem; upon investigation, they found that it was a widespread problem. Subsequent investigation revealed problems with materials, design, manufacture, and installation of the device. Had the NRC better defined the scope and frequency of analysis to locate such a safety problem, it could have been detected and dealt with earlier. The GAO recommended that the NRC better define the scope and frequency of analysis of licensee event reports, and it urged the NRC to establish a system to control and evaluate incident reports with clearly defined objectives. Essentially, the GAO told the NRC to figure out how to use the available information more effectively so that problems can be identified, diagnosed, and dealt with quickly.

The Three Mile Island Commission pointed out room for improvement in developing and disseminating information. Its report notes a lack of closure in the system—that is, important safety issues are frequently raised and may be studied to some degree in depth, but are not carried through to resolution. The lessons learned from these studies do not reach those individuals and agencies that most need to know about them. The commission said that although much information is generated by and available to the NRC, there is no systematic evaluation of it or attempt to find danger signals that could help avert the kind of accident that occurred at Three Mile Island. To correct the situation, the TMI Commission urged the formation of an effective, industrywide communications network to "facilitate the speedy flow of this information to affected parties."[7] In addition to NRC evaluations, the industry should set safety standards and police itself as well.

THE NRC AS REGULATOR: HOW TOUGH—HOW FAIR?

It is not uncommon for regulators to want the business they regulate to succeed, nor is that particularly sinister. For example, the Department of Agriculture, with its numerous regulatory powers, wants farmers to prosper. The multitude of other regulatory and administrative bodies that oversee various segments of the economy must administer the laws and promulgate the rules needed to make the activities of those they regulate fair and safe to the public. Regulators must seek to attain a balance of some sort between fairness and safety to the public and the economic well-being of the industries they regulate. The Nuclear Regulatory Commission should be judged on these terms; but the public should be able to expect that, in view of the potential danger of a nuclear-power-plant accident, if the NRC does err, it will be on the side of ensuring public safety, and that the enforcement of its regulations will be fair but vigorous. Another GAO report suggests that the NRC has not been tough enough on the nuclear power industry.[8]

The NRC has complete jurisdiction over those who handle nuclear materials (plutonium or uranium), source materials (things used to produce nuclear materials), and by-product materials. A license from the NRC is required to handle any of these materials, and this license requires adherence to many regulations. The 1954 Atomic Energy Act states that regulation of nuclear energy must provide adequate protection to the health and safety of the public, and leaves to the regulating agency how to do that. Title 10 of the Code of Federal Regulation enumerates all the regulations governing nuclear power and includes standards for protection against radiation and general design criteria for nuclear power plants. In an attempt to standardize design requirements in the mid-1970s, the NRC held four hearings to establish standards for emergency core-cooling systems, power-plant effluents, the uranium-fuel cycle, and the transportation of fuel and wastes. The generic hearing approach was criticized for not allowing adequate testimony from local interveners or the public, but it had the virtue of moving in the direction of better and more uniform safety systems in nuclear power plants. Assuming that the rules and regulations are adequate to provide reasonable assurance that public safety is being protected (not a universally held assumption), the questions arise, What tools does the NRC have to enforce the regulations? and How willing is it to use them?

When the NRC detects a violation of its regulations, it can do one of three things: send the licensee a letter notifying the party of the violation; assess a civil monetary penalty, that is, a fine; or issue an order directing the licensee to cease an unsafe practice or have their license suspended, modified, or revoked. The third measure can amount to closing a power plant. The General Accounting Office studied actions in 1977 of the NRC and found that there were violations in 40 percent of its 6,512 inspections, about half of which were nuclear power plants. Almost all of the violations, 99 percent, simply resulted in letters to the violators. According to the GAO report, 13 were civil penalties and 5 were orders to cease and desist. For civil penalties, which were the focus of the GAO report, the NRC was limited to $5,000 for each violation and $25,000 for all violations within a 30-day period. In the summer of 1980, the 96th Congress passed a law that raised the maximum civil penalty per violation to $100,000 and eliminated the cap that limited the total penalties payable in a 30-day period. This change provides greater incentive to utilities that run nuclear reactors to err on the side of safety. Still, when power must be purchased for $800,000 a day, as was the case at Three Mile Island, a management decision to close down a nuclear power plant to fix a small irregularity could conceivably be viewed as bad business judgment to the point of money-wasting sentimentality. Even with the higher penalties, the decision to keep a facility running and risk a fine could appear economically attractive in view of the cost of purchased power.

Even assuming that the NRC now has adequate tools to do the job, how aggressively did it use the weak ones it had in the past? The GAO report does not offer encouraging evidence of the NRC's regulatory toughness. In looking carefully at 18 of the 68 civil-penalty cases between 1971 and 1977, the GAO found that the regulatory body, whether it was the Atomic Energy Commission or the NRC, erred in favor of protecting those being regulated rather than the public. In chapter 3 of a recent GAO report, several points were made.[9] When the NRC finds that a licensee has violated a regulatory requirement on separate occasions, or on a continuing basis, it usually cites the licensee for only one violation. This practice reduces civil-penalty amounts and understates the number and frequency of violations found in inspections. In half of the cases investigated, NRC inspectors found that licensees had violated the same regulatory requirement more than once since the previous inspection, but in all but one of those cases the NRC cited the licensee for only one violation. In one case an NRC regional office proposed a civil penalty against a utility because of two personnel overexposures in early 1977, an increase in the frequency of overexposures and weaknesses in the licensee's radiation-safety training program. Realizing that the two overexposures were separable, the regional officials still recommended a single civil penalty of $4,000 for the two overexposures, the minimum amount possible.

In another case, a utility failed to analyze its reactor's cooling water for radioactivity for six months, although the NRC requires monthly analysis. Instead of assessing a fine of $6,000 for the six months, the NRC assessed a fine of $1,000, the authorized fine for just one month.

Strong evidence in the GAO report also indicated that the NRC had not been aggressive in selecting and imposing civil penalties, and tough recommendations for inspectors were often weakened by officials higher up in the regulatory agency. Another example will illustrate this phenomenon. In April 1975 an NRC regional office recommended a $19,000 civil penalty based on 18 alleged violations of NRC's 18 quality-assurance requirements for operating a facility safely. The regional office felt that the violations represented a breakdown in the plant's quality-assurance program. After a lengthy investigation, both the regional and headquarters staffs agreed that 15 or the 18 alleged violations were supportable. The regional staff wanted to assess a $30,000 civil penalty for the 15 violations; but the headquarters wanted to consolidate each violation of any one of the 18 quality-assurance criteria into a single violation for that criterion, reducing from 15 to 8 the number of violations. In the end, the NRC headquarters staff decided that an enforcement letter was better than imposing any kind of civil penalty, a judgment with which the regional office strongly disagreed. The regional authorities firmly believed that a fine would help to ensure an improvement in the licensee's quality-assurance program, but they finally acquiesced to the NRC headquarters decision because of the time (seven months) that had elapsed in trying to resolve the case.

The GAO report also criticized the NRC for withdrawing civil penalties because of lack of clarity in the regulations and then failing to clarify those regulations. Moreover, it urged the NRC to expedite its imposition of civil penalties on utilities building or operating nuclear power plants so that the time between inspection and fine could be cut from 68 days to 40 days.

In its conclusion, the GAO report said that "NRC civil penalties do not provide these licensees economic incentives to improve the safety of their operations, nor do they promote NRC's desired image of a tough but fair regulator." Later it adds that "our review of 22 proposed and actual civil penalty cases leads us to conclude that NRC has not fully and effectively used the civil penalty authority it now has." This report was issues two days before the accident at Three Mile Island, an abnormal occurrence that illustrated in bold relief the need for tough enforcement of NRC regulations.

The NRC's tepid administration of its enforcement powers changed somewhat in early April 1980. At that time the NRC proposed a $100,000 fine against Babcock and Wilcox Company, designer of the Three Mile Island nuclear plant, for failure to report safety information that could have prevented the accident. The proposed fine was the first attempted civil penalty by the NRC against a reactor supplier. Yet the investigation and report that grew out of the accident at Three Mile Island was stinging in its criticism of the NRC. The President's Commission on the Accident at Three Mile Island recommended that the NRC be abolished in its current form. Stating that it had seen evidence that some of the old promotional philosophy still influences the regulatory practices of the NRC, the commission cited managerial problems that start at the very top of the NRC. They advocated a new agency to be headed by a single administrator (to be selected from outside the agency) so that accountability and organizational efficiency could be enhanced, a recommendation that President Carter rejected. The current NRC, with five leaders, practically ensures that no one is really in charge.

The NRC would have a big job to do if it were to make the substantive changes that the President's commission suggested. In addition to improving operator training and information flow, the reorganized NRC would have to reorient the mind-set of both government regulators and the utilities in two important ways: the safety program would have to be geared as much to the small loss-of-coolant accident like that at Three Mile Island as it is now to the accident that involves major parts of the reactor; and the safety program must be taken out of the quagmire of written rules and regulations that burden government officials but do little to ensure the safety of the power plant. The new agency would have to better establish what parts of the reactor are safety-related; the commission's report put it in the following way: "At TMI-2, the PORV [pilot-operated relief valve] was not a 'safety-related' item because it had a block valve behind it. On the other hand, the block valve was not 'safety-related' because it had a PORV in front of it." (In view of what happened at Three Mile Island, one could almost make the case that everything in a nuclear power plant should be considered safety-related.) If a new agency ever takes the place of the NRC, it might best start from scratch in setting up a strategy of regulation, using the report of the President's Commission on the Accident at Three Mile Island as its starting point.

Bearing in mind the questionable training of operators in accident management, the shortcomings of the information system used to help prevent accidents, the weakness of the NRC as an effective regulator, and the unexpected chain of events that occurred at Three Mile Island, let us turn now to attempts at risk assessment as they reflect on the safety of nuclear power plants. Working backward, perhaps, it has been shown that accidents of a very serious nature do occur; now we can look at efforts to predict whether, according to mathematical probability calculations, they should occur.

CATASTROPHES: THEY ARE NOT SUPPOSED TO HAPPEN

Risk assessment traditionally looks both at the likelihood of an accident occurring and the potential outcome of that accident. Advocates of nuclear power usually emphasize the remoteness of the odds of a full-blown calamity, and opponents stress how bad it would be if such a calamity occurred. Assessing the risk of nuclear power plants is more difficult than assessing other kinds of risks, such as the risk of having an automobile accident, being injured by a fall, or drowning. In the case of nuclear power plants, the technology is too new and the number of plants too few to provide any kind of solid, actuarial basis for estimating the risk of a dangerous meltdown.

Thus, those who try to figure out the likelihood of a nuclear-power-plant catastrophe are at a disadvantage. They have inadequate "real-world" bases for their projections; they only have mathematical estimates based on theory rather than experience. Moreover, unlike other technologies, such as aircraft, for example, woefully little has been done to provide nuclear test pilots or test facilities that span the life expectancy of a commercial nuclear power plant. The result has been that the bugs have had to be worked out at the same time that full-scale commercial nuclear power was developing, an unfortunate fact when one attempts to assess the true safety of nuclear power as measured by experience rather than theory.

An example of the learning-by-doing philosophy of the nuclear power industry can be found in a malfunction that occurred at Florida's Crystal River nuclear plant in early March 1980, nearly a year after TMI. At the Crystal River plant, the same design flaw that created the problem at TMI caused the same valve to stick open. At Crystal River, however, an operator who had become more watchful because of TMI, quickly noticed the open valve and blocked it with a backup valve, thus averting the chain of events that occurred in Pennsylvania. With a technology as potentially dangerous as commercial nuclear power, one could challenge the responsible thinking behind the decision to rely so heavily on the learning-by-doing approach that is evident here.

Accepting, then, that risk assessment for nuclear power cannot be adequately grounded in experience, and that such assessment must look both at the theoretical likelihood of an accident and its theoretical outcomes, one more element needs to be added in evaluating nuclear power's desirability from the standpoint of safety.

This involves the idea of necessary risk. The necessity of taking a risk is determined by the need of the consumers for the product and the alternatives available to meet consumer needs, when this concept is viewed in a commercial context.

For example, U.S. consumers feel they have a strong need for mobility. They want to be able to travel to stores, football games, the countryside, or anyplace else whenever they wish. So far the primary technology for doing that is the automobile. There is a risk in driving a car: every year more than 50,000 Americans die in automobile accidents. But if the United States is going to enjoy the level of mobility that it does, the risk of death is a necessary risk, since the automobile is the primary available way to achieve such mobility. Even flying in airplanes can be seen as a necessary risk in the twentieth century United States. Often a person's job or recreation needs are such that unwillingness to fly cuts one off from career advancement or enriching travel experiences. As long as there is a strong need to travel long distances fast, airplanes will be a necessary risk, since they are currently the only way of doing it.

Everybody needs a certain amount of electricity. Those with even the most austere lifestyle benefit from electricity in countless ways; it is important to our lifestyle. However, the amount used and the way it is generated varies. The next chapter will look at alternatives to nuclear power in an effort to show that this particular technology poses an unnecessary risk, since the same or similar consumer needs can be met by means of other technologies. The concept of necessary risk is put forward here as an attempt to give greater depth to the issue of risk assessment.

THE RASMUSSEN REPORT AND THE LEWIS REPORT

The most ambitious and expensive effort to assess the risk of nuclear power plants came in the form of the so-called Rasmussen Report, which was undertaken in 1972 and published in 1975. Led by Norman Rasmussen, professor of nuclear engineering at the Massachusetts Institute of Technology, the study attempted to provide an overall assessment of risk based on the probabilities of failure of various parts of a nuclear power plant, operator error, poor maintenance, natural hazards, and other causes. Using techniques developed over the preceding ten years by the Department of Defense and the National Aeronautics and Space Administration, the study was based on the use of "event trees" and "fault trees." An <u>event</u>

tree defines an initial failure within the plant and traces the possible events within its various components that could lead to meltdown and radiation release. According to the report, "Event trees were used in this study to define thousands of potential accident paths which were examined to determine their likelihood of occurrence and the amount of radioactivity that they might release." A fault tree attempts to assess the likelihood of failure of the various systems in the plant, based on data covering the components, such as pumps, pipes, and so forth, as well as the likelihood of operator and maintenance error.

Three years and $4 million after the project began, the news was very good for nuclear power advocates. According to the report, the chances of an individual being killed by a nuclear-power-plant accident were 1 in 5 billion, compared with 1 in 4,000 for car accidents. The probability of 100 or more people being killed by a reactor accident (assuming 100 reactors in operation) was the same as the likelihood of their dying from a meteor impact. All the figures coming out of the Rasmussen Report were equally reassuring about the safety of nuclear power plants. However, not everybody was satisfied with the report.

Employing the same techniques used by the Rasmussen Report, the Sierra Club and the Union of Concerned Scientists (UCS) evaluated the probability of an accident that had actually occurred in 1970 at the Dresden nuclear plant in Illinois.[10] The result of their calculation was an accident-probability estimate of one in one billion-billion. Essentially, they had discovered that the accident that had occurred was not supposed to, according to the kind of complex and sophisticated mathematical calculations used by the Rasmussen Report; after a certain point the odds become remote enough to make the possibility of an accident negligible. The calculations done by the Sierra Club and the UCS revealed, among other things, the inadequacies of a prediction system grounded so little in actual experience.

The Rasmussen Report became controversial enough to cause the federal government to reassess its findings. On July 1, 1977, the NRC formulated the Risk Assessment Review Group, headed by H. W. Lewis, professor of physics at the University of California, Santa Barbara.[11] The Lewis Report, published in September 1978, had many serious criticisms of the Rasmussen Report, although it termed the report an improvement over previous attempts at risk assessment. According to the Lewis Report, the shortcomings of the Rasmussen Report included the following:

1. Understatement of the error bounds of its probability estimates. The Lewis Report said: "This is true in part because there is in many cases an inadequate data base, in part because of an inability to quantify common cause failures, and in part because of some questionable methodological and statistical procedures."[12]

2. Failure to give adequate weight to certain initiating events. The Lewis Report indicated that it was not convinced that such things as fires, earthquakes, and human-accident initiation deserve to be given as little attention and be viewed as negligibly threatening as the Rasmussen Report suggests they should be.

3. Inadequate statistical analysis. The Rasmussen Report suffers from "lack of data on which to base input distributions to the invention and use of wrong statistical methods. Even when the analysis is done correctly, it is often presented in so murky a way as to be very hard to decipher."[13]

4. Poor peer review. The Lewis Report indicated the need for first-class peer review to justify confidence in such a comprehensive study, and adequate peer review was lacking.

5. The misleading nature of the executive summary. This document, the most widely read part of the Rasmussen Report, "does not adequately indicate the full extent of the consequences of reactor accidents; and does not sufficiently emphasize the uncertainties involved in the calculation of their probability."[14]

The events at Three Mile Island six months after the Lewis Report demonstrated the inherent limitations of any risk-assessment exercise done with too little foundation in experience. The numerical specificity of the kind of catastrophe prediction found in the Rasmussen Report gives the illusion of certainty and accuracy in a way that ultimately is a disservice to the public. It is a use of statistics that is as misleading as body counts proved to be in the Vietnam War. According to studies like the Rasmussen Report, the danger of nuclear power plants is negligible. This conclusion is grounded in sophisticated mathematics, computer use, and statistics, but is not adequately grounded in experience, traditionally the best guide of all.

CONCLUSIONS AND COMMENTS

How safe are commercial nuclear reactors? There is certainly no way to quantify an answer to that question. In a way, much of the basis for any answer is related to intangible and imponderable issues. What does one want to believe? I prefer to believe that there will be no full-blown catastrophe at either of the two nuclear power plants near the Twin Cities, where I live, and that the redundant safety features will ensure that my family, friends, and the public will be safe. What I have read, however, does not encourage me about the safety of nuclear power or the system that operates and oversees it. Briefly, here is why.

1. Nuclear power plants are extremely complex, and a minor malfunction can lead to a chain of events both unanticipated and potentially calamitous.
2. The ability to detect, identify, and respond to malfunctions is less developed than I had thought.
3. The communication within the nuclear power industry regarding a defective or potentially defective mechanism is much worse than I would have thought.
4. The quality of operator training is uneven and not adequately oriented to managing an accident correctly.
5. The cost of buying electrical power when a nuclear power plant is closed is so high that management could well decide against closing a facility just to repair a minor part whose malfunction would have no apparent impact on plant safety.
6. The NRC has been demonstrably lax in enforcing regulations, and in the past it lacked adequate enforcement tools, facts that were pointed out by an agency of the U.S. government, not nuclear power critics.

To be sure, much of the time nuclear power plants work the way they are supposed to. However, this is an unforgiving technology in the sense that one major accident could injure or kill thousands of people. In that regard, what is required of it from the standpoint of safety is a level of technical perfection that I, for one, believe is unreasonable. A friend of mine asked me why we could not apply the same level of effort and technological precision to keep nuclear power plants safe as we do to ensure a successful space shot. Even disregarding the deaths of the three astronauts in the accident that aborted one of our space probes, there are two crucial differences between the two kinds of technological endeavors: a worst-case scenario for a nuclear power plant will kill more than three persons; and there are 72 commercial nuclear power plants and over 90 additional ones on the way, making attempts at accident prevention correspondingly less focused.

No technology is risk-free. Someone trying to sort out the issues involved in the nuclear power controversy must try to weigh the varying risks and benefits of nuclear power and its alternatives. Our nation will continue to have very real needs for electricity, and, before a judgment on nuclear power can be formulated, some analysis of the options must be made.

As a politician trying to pass a bill through the Minnesota State Legislature, I wanted to show other members the alternatives to nuclear power. For conservative members particularly, some of whom were farmers or small businessmen and whose need for reliable electricity was immediate and strong, I needed to make the case that alternatives to nuclear power did exist and were attainable today. Even though my bill would not have closed down existing nuclear power facilities, any curtailment of electricity even in the future was a direct threat to some of the members of the House of Representatives and the people who had sent them to St. Paul to represent them.

How warranted is the skepticism about alternatives to nuclear power, and how necessary is the risk of this controversial technology?

NOTES

1. Winthrop Rockwell, "Three Mile Island Revisited," Corporate Report, March 1980, p. 32.
2. Ibid., p. 33.
3. Jim Dawson, "N-Training Plant: Violations?" Minneapolis Star, May 9, 1979, p. 1.
4. President's Commission on the Accident at Three Mile Island, Report of the President's Commission on the Accident at Three Mile Island (Washington, D.C.: U.S. Government Printing Office, October 30, 1979), p. 68.
5. Ibid., p. 23.
6. U.S., General Accounting Office, Reporting Unscheduled Events at Commercial Nuclear Facilities: Opportunities to Improve NRC Oversight (Washington, D.C.: Government Printing Office, January 26, 1979), pp. 1-21.
7. President's Commission on the Accident at Three Mile Island, Report, p. 68.
8. U.S., General Accounting Office, Higher Penalties Could Deter Violations of Nuclear Regulations, Washington, D.C., February 1979, pp. 1-25.

9. U.S., General Accounting Office, NRC Has Not Effectively Used Its Civil Penalty Authority (Washington, D.C.: Government Printing Office, March 1979).

10. Ralph Nadar and John Abbotts, The Menace of Atomic Energy (New York: W. W. Norton, 1977).

11. U.S., Nuclear Regulatory Commission, Risk Assessment Review Group to the U.S. Nuclear Regulatory Commission, NUREG/CR-0400, Washington, D.C., September 1978.

12. Ibid., p. viii.

13. Ibid., p. ix.

14. Ibid.

REFERENCES

Articles

Dawson, Jim. "N-Training Plant: Violations?" Minneapolis Star, May 9, 1979, p. 1.

Lanouette, William. "Kemeny Panel Will Draw Blueprint for Future of Nuclear Power." National Journal, August 18, 1979, pp. 1366-68.

Morland, Howard. "The Meltdown That Didn't Happen." Harpers, October 1979, pp. 16-27.

"The Next Three Mile Island." New York Times, November 4, 1979, p. 20E.

Rockwell, Winthrop. "Three Mile Island Revisited." Corporate Report, March 1980, pp. 32-33.

"Speak from the Heart on Nuclear Power." New York Times, October 2, 1979, p. A-22.

von Hippel, Frank. "Looking Back on the Rasmussen Report." Bulletin of Atomic Scientists, February 1977, pp. 42-47.

Reports

Congressional Research. Nuclear Power: Safeguards. Issue Brief no. IB 78212, reprinted in booklet form.

McCracken, Samuel. "The War against the Atom." Commentary, Boston University, 1979.

Michelson, Carlyle. Decay Heat Removal during a Very Small Break LOCA for a B&W-Fuel-Assembly PWR. Report prepared for Babcock and Wilcox. Knoxville: Tennessee Valley Authority, January 1978.

National Council of the Churches of Christ, Committee of Inquiry. "Background Report to the National Council of Churches of Christ in the USA in Support of the Plutonium Economy: A Statement of Concern." September 1975.

Pollard, Robert D. "The Nugget File." Report prepared for the Union of Concerned Scientists, January 1979.

President's Commission on the Accident at Three Mile Island. Report of the President's Commission on the Accident at Three Mile Island. Washington, D.C.: U.S. Government Printing Office, October 30, 1979.

Union of Concerned Scientists. "To the Brink of the Abyss: The First Hours of Three Mile Island." Nucleus (newsletter of Union of Concerned Scientists), 1979.

U.S., General Accounting Office. Reporting Unscheduled Events at Commercial Nuclear Facilities: Opportunities to Improve NRC Oversight. Washington, D.C.: Government Printing Office, January 26, 1979.

――――. Higher Penalties Could Deter Violations of Nuclear Regulations. Washington, D.C., February 1979.

――――. NRC Has Not Effectively Used Its Civil Penalty Authority. Washington, D.C., March 1979.

U.S., Nuclear Regulatory Commission. List of Abnormal Occurrences at Nuclear Power Plants. Washington, D.C., July 1979.

――――. Reactor Safety Study: An Assessment of Accident Risks in US Commercial Nuclear Power Plants. WASH-1400, Washington, D.C., October 1975.

――――. Risk Assessment Review Group to the U.S. Nuclear Regulatory Commission. NUREG/CR-0400, Washington, D.C., September 1978.

Books

Fuller, John G. We Almost Lost Detroit. New York: Reader's Digest Press, 1975.

Nader, Ralph, and John Abbotts. The Menace of Atomic Energy. New York: W.W. Norton, 1977.

5

HOPEFUL ALTERNATIVES: THE FUTURE IS NOW

Before I became involved in nuclear-waste legislation, I had believed that coal was the only realistic alternative to nuclear power, and that bothered me. Trading radiation for acid rain did not strike me as a great bargain. My perception that the choice was between coal and nuclear power also made me a little suspicious of those who protested visibly against nuclear power. I felt their cause might simply be another fashionable issue for angry liberals, with a certain generalized antiestablishment motivation behind it. What would the same people be protesting next year? More to the point, What did they advocate as an alternative to nuclear power? I did not see marches or demonstrations for positive, realistic alternatives. They were not carrying More Coal Now signs. I have always been a bit distrustful of trendy movements, and the antinuclear cause was no exception. What were the alternatives?

In addition to my belief that coal was the only realistic alternative to nuclear power, I had several other preconceptions about the alternatives.

1. If nuclear power is phased out, some other large-scale centralized system must take its place, such as nuclear fusion.
2. Much more research is needed before other means of generating electricity will be feasible.
3. The idea of generating electricity by the old methods, such as windmills, is a sentimental throwback to an age gone by; no sig-

nificant contribution to current needs for electrical power can be made by such outdated technology.

4. The electrical power industry knows how to accurately predict increases in demand for electricity, thus minimizing the risk of expensive overbuilding of generating capacity.

5. There is no realistic strategy for phasing out nuclear power in the near future.

If nuclear power were to be phased out in this country, obviously something would have to take its place. Any alternative means of meeting our electrical needs should compare favorably with regard to cost and safety and should be attainable now or very soon. How economically competitive the new technology or technologies are will be colored by the degree to which the cost of nuclear power continues to rise; based on the experience of the 1970s, further escalation seems likely. In some cases, the alternatives, in order for their costs to be fairly compared with that of nuclear power, should be able to benefit from the same kind of federal largess that has propped up the technology of nuclear fission.

Whatever replaces it should provide less risk to the public safety than does commercial nuclear power. A distinction needs to be made here between risk to the public, which has been fundamentally voiceless in the decision to expand a technology such as nuclear power, and the kind of private risk that accompanies the use of a different, more-decentralized technology to meet the same needs. For example, somebody may be seriously injured or die in attempting to install a wind generator at a farm or urban home, which could prompt the proponents of nuclear power to comment that such an accident hurt more people than did Three Mile Island. The critical point is that risks in alternative, decentralized technologies are consciously taken by the consumer, not forced on them by a utility. As I will show below, the most important alternative to nuclear fission is as risk free as any technology can be.

The alternatives to nuclear power should be attainable now or in the near future. Nuclear fusion, for example, may someday play an important part in meeting the nation's energy needs, but it is still in the embryonic stage of development and, at best, will not be available until well into the twenty-first century. Alternatives need to be found now.

THE TECHNOLOGY OF CONSERVATION AND ENERGY EFFICIENCY

All other things being equal, a new technology is better than an existing one if it does a given job more efficiently. Efficiency is a measure of how much work must be done to achieve a desired result: the less work required for the same result, the more efficient the technology. Because it promises the most efficient alternative to nuclear power, conservation merits the most attention of all the competing alternatives. The key to conservation is not "doing without"; rather, it is finding ways to eliminate waste and, as in the case of electricity, finding ways to use it more intelligently. A look at some examples of successful electricity conservation will show how realistic and immediately available it is. First, however, brief comments on the use and pricing of electricity are in order to provide an adequate backdrop to meaningful discussion of conservation.

In looking at the use of electricity, one should bear in mind that two-thirds of the energy used in producing traditional central-station electricity is wasted. From an energy efficiency point of view, then, electricity is truly a premium product that should only be used when other, less wasteful alternatives do not exist. Amory Lovins, energy expert and writer, has put the case cogently.[1]

Lovins pointed out that 58 percent of all energy at the point of end use is required as heat and another 38 percent is used for mechanical motion (31 percent for vehicles, 3 percent in pipelines, and 4 percent in industrial motors). The remaining 4 percent is used for such purposes as lighting, telecommunications, and electrometallurgy, which require electricity. Thus, although in 1976 the nation used electricity for 13 percent of its end-use needs, only 8 percent of such end uses required electricity. Given that two-thirds of the heat used to generate electricity is wasted, it would seem desirable to reduce the gap between 13 percent and 8 percent. Lovins suggests that we could even go further.

> So limited are the U.S. end uses that really require electricity that by applying careful technical fixes to them we could reduce their 8 percent total to about 5 percent (mainly by reducing commercial overlighting), whereupon we could probably cover all those needs with present U.S. hydroelectric capacity, plus cogeneration capacity available in the mid-to-late 1980s. Thus an affluent industrial economy could advantageously operate with

no central power stations at all! In practice we would
not necessarily want to go that far, at least not for a
long time; but the possibility illustrates how far we are
from supplying energy only in the quality needed for the
task at hand.[2]

Lovins' analysis provides a useful framework for assessing the potential of conserving electricity and affirms the need for developing a system that provides incentives to do so.

The fact is that the way electricity is now priced and regulated, there are inadequate incentives to conserve electricity. This is a point forcefully made in a study by Edward Berlin, Charles Cicchetti, and William Gillen and overseen by S. David Freeman, now the head of the Tennessee Valley Authority. The findings of that study are most revealing.[3]

There is no reason to believe that electricity is any different from other products in one important way: as price goes up, demand should go down, or in other words, electricity is cost sensitive. That is a crucial premise to the discussion that follows. When a power plant is proposed and built, its generating capacity is shaped by the expected peak demand. So, for example, if for 20 hours of the day the demand is X, but from 4:00 P.M. to 8:00 P.M. it is X plus Y, the generating capacity must be high enough to handle X plus Y, the peak demand. That means building a bigger power plant than would be needed if that same peak could be spread around in a way that would reduce the amount of electricity demanded at any one time.

If there were a way to persuade people and businesses to use their electricity at nonpeak hours, the total generating capacity could be reduced, and utility bills could be kept lower. The best way to achieve that goal would be to make it more expensive for the consumer to use electricity during peak hours. In other words, the pricing of electricity should be changed. Currently, a kilowatt is often priced the same no matter what time of day it is consumed: electricity used at peak-demand time is priced the same as if it were used at three o'clock in the morning. For most products, as demand goes up, prices go up, and this causes a reallocation of precious resources. But electricity is often priced without consideration of the effects of the cost of generating at peak capacity. The current system provides inadequate incentive for users of electricity to hold back at the busiest hours.

Another unfortunate effect of the current pricing of electricity is that utility companies often can and do give volume discounts to large customers who consume the greatest amount of electricity. In other words, the more they use, the less expensive per kilowatt it becomes for them. Rather than creating an incentive to conserve electricity, this pricing system rewards those who consume the most electricity by qualifying them for volume discounts.

Both heavy use of electricity at times of peak demand and wasteful overconsumption by large customers would be corrected if electrical pricing were to change. The way pricing occurs if the market functions perfectly ensures that the price of a product equals the marginal cost to the business producing it. In a regulated monopoly, however, no such pure economic theory can be expected to hold. Some of the benefits of marginal cost pricing could occur, however, if a system of peak-load pricing were to be instituted.[4] Peak-load pricing simply allows the price of electricity to fluctuate according to the demand for electricity during the day, so that it would cost more at times of peak demand and less at nonpeak hours. It partially subjects electricity to the same kind of market forces that most products must respond to and more accurately reflects its true cost at any given time. Peak-load pricing is an idea that is receiving considerable attention throughout the utility-regulating community. Its rapid implementation in some form will be an important component of any successful energy-efficiency program.

Peak-load pricing will cause a significant reduction in the disparity between peak demand and nonpeak demand, which will in turn reduce the demand for large generating facilities. A lower peak demand serving the same population will mean either smaller power plants or longer periods between the construction of larger ones. It will make the use of electricity more efficient. In the context of the debate on nuclear power, it provides an important tool for curbing the growth of all electrical power generation, including what is generated by commercial nuclear power plants.

Over half the states have time-of-day pricing for commercial and industrial users of electricity, and there is a growing number of states that have it in the residential sector as well.[5] The federal government is encouraging states to move in this direction; under the Public Utility Regulatory Policy Act of 1978, states are required to hold hearings on this issue. Utility companies appear increasingly receptive to the idea, although they point out that one problem with time-of-day pricing is the cost of installing an additional meter to make the necessary readings, a cost more burdensome for residential consumers than for businesses.

In addition to peak-load, or time-of-day, pricing, there is the issue of whether to structure electrical rates to reward overall conservation efforts. Proponents of so-called life-line electrical pricing urge adoption of a system that charges the customer less per kilowatt up to a certain level of consumption, and then more beyond that level. It is clear that if greater consumption will cause consumers to pay more per kilowatt, they will have a clear incentive to conserve. Legislatures that value energy conservation will require public service commissions to restructure rates in this manner.

Regardless of when and how the needed regulatory and pricing changes take place, there are many examples of imaginative electricity conservation; two models illustrate ways of reducing demand for electricity.

Los Angeles

Since half of Los Angeles's electricity is generated by oil, the 1973 Arab oil embargo threatened the city's electrical power system, which is run by the Department of Water and Power, a public agency. More than 11 million barrels of oil that had been contracted for were held up, leading the press and others to speculate on what was going to happen when the lights went out. City officials contemplated limiting the work week, instituting neighborhood-by-neighborhood blackouts, and drastically raising the prices. They dismissed all three options as being unworkable. Instead, an ad hoc committee of civic, business, and labor leaders came up with the idea of setting mandatory targets for reduction of consumption for all customers, but leaving to those customers the decisions on how to meet the goals.

Under phase 1 of the plan, customers were to cut back on their use of electricity compared with the same billing period of the previous year; failure to comply would result in a 50 percent surcharge on their bills. This was intended to reduce the city's consumption by 12 percent, and this reduction was to be phase 2, which would have set even higher targets. However, phase 1 was so successful that phase 2 was not needed. Here are the results of phase 1 (in percent).

	Target	Actual Reduction
Residential	10	18
Industrial	10	11
Commercial	20	28

Source: Robert Stobaugh and Daniel Yergin, eds., Energy Future (New York: Random House, 1979), p. 145.

The total reduction was 18 percent, as opposed to the target 12 percent. After the oil embargo was lifted, the program was suspended in May 1975; but even a year later electricity sales were 8 percent lower than they had been in 1973.

Although the Department of Power and Water provided strong incentives to the customers to comply with the targets, much of the success of the Los Angeles program was due to the consensus voluntarily arrived at by various affected groups as to how the program should work. Furthermore, discretion was left to the consumers on how and where they would cut their consumption. That the total demand for electricity remained lower in 1975 than it had been before the program began is a testament to the potential for both cost and energy savings of an aggressive conservation effort. While the Los Angeles experience did not take place in the context of the nuclear power controversy, this example is of obvious importance when assessing alternatives to nuclear power.

Seattle

Of all the major cities in the United States, Seattle has the least expensive electricity, owing in large part to its heavy reliance on hydroelectric power. However, new additional power would be considerably more expensive. In 1975 Seattle City Light, another municipally owned utility, was considering buying future power from two proposed nuclear power plants that were to serve the Washington Public Power Supply System, a consortium of some 20 such utilities. Objections to that proposed way of meeting Seattle's energy needs on the part of the local citizenry caused the city council to reassess the issue.

A broad-based citizens group was formed to look at the long-range electrical needs of Seattle and to lay out alternative ways to meeting those needs. The city council decided on conservation as the preferable technology for meeting the city's needs and reached the judgment that nuclear power was the least desirable alternative. The council agreed on a plan such that by 1990 there would be a 230-megawatt reduction in the proposed consumption of electricity, 145 megawatts coming from commercial and industrial concerns and 85 megawatts from residential consumers. By the spring of 1979, consumers of all kinds had already cut nearly 70 megawatts from the projected growth in demand, well on schedule.

Because electricity costs Seattle residents only 1.8 cents per kilowatt-hour, it would have been difficult to bring about conservation through the pricing system. Instead, an aggressive public relations program aimed at educating business persons and residents of Seattle, coupled with city ordinances, has brought about the desired results. Seminars are put on by Seattle City Light for homeowners, hospitals, schools, supermarkets, nursing homes, and other government agencies to train people how to use their electricity more wisely. A commercial energy code for new buildings and tough service requirements for residential dwellings that are to be converted to electric heat are also helping the conservation effort.

Strong objection to nuclear power figured prominently in the city government's overall reassessment of how to meet future demand for electricity and led to a rational, planned means of moving toward a more acceptable alternative. It is to be hoped that investor-owned utilities will be equally open to citizen concerns and equally committed to finding alternatives to nuclear power. Seattle showed that even in a city with extremely inexpensive electricity, effective conservation is possible. The impetus for change when considerable cost savings are at stake would appear to be even more compelling in other cities.

Some Technologies of Conservation

In addition to electricity-conservation programs that have been shown to work, there are a number of conservation ideas that have yet to be exploited generally but are equally promising.

One such idea has been offered in a publication put out by the Massachusetts Institute of Technology. An article written by Marc H. Ross and Robert H. Williams points out the considerable savings that are possible in the wiser use of air conditioning of high-rise office buildings.[6] Using a hypothetical ten-story, million-square-foot office building in New York City, the authors found that only one-sixth of the air-conditioning load is due to heat conduction from the outside and solar radiation through the windows. Over one-half of the load is due to heat generated by lighting (about 6 watts per square foot) and about one-fifth is due to ventilation (20 cubic feet per minute per person). In this example it is considered reasonable to reduce illumination levels to 1.5 watts per square foot and ventilation to five cubic feet per minute per person. The combination would reduce the air-conditioning load more than 50 percent (more ventilation would be required in buildings that allow smoking everywhere). If heat exchangers were used to recover "cool" from exhausted air, the reduction in air-conditioning load could be raised to 70 percent.

A key point Ross and Williams make is that lighting within an office building is a far greater source of heat in the summer than are the hot summer air and the sunlight. That fact should figure prominently in future energy planning for new buildings, as well as efforts to adjust current lighting systems to be more energy efficient. As for ventilation, the requirements that exist in buildings that fail to provide no-smoking areas are such that excessive electricity is consumed to circulate the office air appropriately (demand for electricity could go down if fewer people smoked, or if more of the office space were designated no smoking.

More recycling of aluminum is another way to conserve electricity. Dean Abrahamson asserts that the primary-metals industry is the largest consumer of electricity in all of business.[7] Moreover, aluminum is the biggest user of electricity within that industry; it requires five times more electrical power to produce than does steel. Obviously, then, substituting steel for aluminum whenever practical will do much to conserve electrical power. Even better, the need to recycle aluminum takes on a dimension that goes beyond better solid-waste management. If the aluminum needs of the United States could be better met through recycling rather than mining and processing, not only could the product be made less expensively with less environmental degradation, but a great deal of electrical energy could be saved.

The malfunction of the chiller turbine of the air-conditioning system at the University of Wisconsin-Milwaukee led to an energy-saving alternative that saved that institution over $200,000 the first year. Instead of using the chiller turbine, 40° F water from Lake Michigan was pumped directly into the air-conditioning system. The necessary adjustment of the valve system only cost $12,000. There were two important prerequisites for the university to make its idea successful: the lake had to be sufficiently cool, which was largely a function of the prevailing winds that summer, and there was an existing pipeline to the lake. Still, any large public or private institution similarly located near large bodies of water, such as the Great Lakes, would do well to investigate the possibilities of following the lead of the University of Wisconsin-Milwaukee, because after the initial investment is made there is a virtually inexhaustible supply of cool water. Such an alternative would not preclude remaining tied in to the conventional system of air conditioning if climatic conditions become unfavorable.

Another idea with industrial application involves hydraulic pumps. Fluids that are pumped for making a product can be driven by a hydraulic pump, an air pump (the most common kind), or a centrifugal pump—all of which need electricity. The paint-finishing industry is a heavy user of pumps, and the painting of new automobiles, for example, requires a sizable expenditure of electrical energy. It requires five times more electricity to compress air than it does to pressurize a column of oil. Graco Company of Minneapolis, which still sells far more air pumps than hydraulic pumps, demonstrated the savings at a recent national sales meeting: for an eight-hour-per-day shift, 250 working days per year, the cost of energy for four hydraulic pumps was $360, as opposed to $1,800 for four air pumps.

Although it is the most energy-wasteful means of heating one's home, electricity is used throughout the United States to do just that. When electricity is used for this purpose, aggressive attempts at home weatherization, insulation, and overall tightening up of the house are crucial to keeping consumption and costs down. According to Amory Lovins, energy lecturer and British representative of Friends of the Earth, electricity should not be used for heating because its real cost is comparable to paying $97 for a barrel of oil. (Lovins has stated that only 8 percent of the nation's energy is nonthermal electricity, which goes to the kind of "premium" use that absolutely requires electricity.) Still, since electricity is sometimes used for home heating, all of the conventional conservation methods available to oil and gas customers should be fully exploited by those who get their heat from electricity.

A device developed both by Exxon and the National Aeronautics and Space Administration, called a power-factor controller, offers much energy-saving potential. This inexpensive device is attached to an electric motor and adjusts the voltage of the motor to match the load. Since electrical motors consume 58 percent of all electrical energy, according to the U.S. Department of Energy, the power-factor controller is a very significant breakthrough. Some estimates are that this device could save up to a British thermal unit (Btu) equivalent of 400 million barrels of oil every year.

Furthermore, the rising cost of electricity has given impetus to increased efficiency in motors. Engineers point out that when electricity was cheap, motor manufacturers skimped on windings, the heart of the motor, in order to save on copper. This lowered the efficiency of the motor. Now that the price of electricity has risen, more costly and efficient motors can save enough money to pay for the increased cost of the motor.

Not smoking at work, recycling aluminum cans, or installing a power-factor controller do not appear to have much to do with nuclear power, but they illustrate the connection between demand for electricity and almost every phase of life, as well as the products that are used to make it more enjoyable. Yet a solitary individual who wants to conserve electricity directly or indirectly in order to reduce the need for nuclear power can make a difference only if he or she is joined by many others doing the same thing. What is needed is an organizational mechanism that can help unite people in a common effort, thus reinforcing their attempts at conservation. The examples of Los Angeles and Seattle are instructive of the possibilities of collective action. Organizing conservation deserves a further look.

Organizing Conservation

The key to success in both Los Angeles and Seattle was the early and constructive involvement of a broad-based group of citizens. In one case the group was formed to respond to an imminent shortage of oil, which was heavily used for generating electricity; in the other case, environmentally conscious and responsible citizens forcefully pushed their case for a reassessment of their city's future dependence on nuclear power. In neither case did the politicians or local government agencies unilaterally declare a policy and attempt to force the citizens to comply with it. The wisdom of adopting a strong conservation program in each city has been borne out by reduced consumption of electricity in Los Angeles and Seattle, with no discontent on the part of the businesses and citizens affected. Political leadership helped ensure the success of each city's program, but the initiative came from the people. If the Los Angeles and Seattle models are to be emulated effectively, the people must be involved from beginning to end, since they are the only ones who can make any conservation program work. One possible vehicle for initiating such a program in a county, city, town, village, or township could be a committee called, for example, Citizens United for Responsible Energy (CURE).

A CURE committee, made up of labor, business, civic, religious, and political leaders, would provide the impetus for a communitywide electrical-energy-conservation campaign. Utility companies could help provide information, as could local energy agencies. The CURE committee could push for laws and ordinances that would encourage conservation of electricity as well as follow the activities of, and testify to, regulatory bodies such as public service commissions. For example, it could serve as a catalyst for a peak-load pricing program. Participation would be open to any interested citizen or business. Crucial to the success of a CURE campaign would be the setting of goals, followed by an accurate assessment of reaching those goals, somewhat similar to how the United Way works in this country. The structure, goals, and modus operandi of each CURE committee would be different according to the special needs of a given community. (The nature of energy issues that could be addressed by a group of this kind would not have to be limited to electrical-energy conservation, but that could be a starting point.)

Why would business, labor, and citizens decide to organize and unite to conserve electricity? Some would be attracted to the environmental benefits of reducing demand for electricity; others would see it as an effective tool to obviate the needs for any new nuclear power and as a means of hastening the transition to a nonnuclear society; but everybody would see one obvious and clear benefit —it would save them money. Money would be saved in two ways: changes made in individual companies, other institutions, and homes would deliver immediate savings to the users of electricity; and the collective success of the CURE committee would greatly slow down or eliminate the need for new electricity-generating facilities of any kind and thus curb rate increases owing to capital expansion.

In the end, needless expansion of electricity-generating power plants is bad for everybody. Energy consultant Amory Lovins makes a strong case that proliferation of power plants, rather than more conservation, is even bad for utility companies.[8]

THE WISDOM OF CONSERVATION FOR UTILITIES

In a paper delivered at the E. F. Hutton Fixed Income Research Conference on Public and Investor Owned Electric Utilities, Lovins made a compelling case for conservation as the best business decision utilities can make today.[9] Ignoring for the moment the needs of the consumer/citizen and the desirability of delivering electricity as efficiently, safely, and inexpensively as possible, and strictly from the standpoint of what is good for the utility, it is arguable that it makes better business sense to curb the growth of power plants that require massive capital outlays. Electrical utilities are the most capital intensive of all energy-producing businesses; more dollars are invested to produce a given usable unit of electrical energy than in any other part of the energy industry. Lovins describes the problem in the following manner:

> Electric utilities combine extraordinary capital intensity with long plant construction times (lead times). In particular, typical lead times are much longer than the time constant of short-run price elasticity of demand. Installed capacity is therefore likely to overshoot its economically sustainable level. To put it in mundane terms, during the long construction period, a utility must generally raise its electricity price in order to finance the construction, keep up its bond ratings and coverage ratio, maintain dividends, and keep common stock values near book. But consumers respond to the higher price (and even to expectations of it) long before the plant is finished, so when it is commissioned, it sells too little electricity for its revenues to cover its fixed charges. (This is inevitable because utilities that assume a nonzero price elasticity compute it on historic costs, not on marginal costs.) The shortfall of revenue makes it necessary to raise prices further, so that the rate of growth—and conceivably (as in Britain) the level—of demand is dampened further—requiring a still higher price to amortize existing plants, and so on into the "spiral of impossibility." For cash flow to collapse, it is not necessary that the actual level of demand should decrease; it is sufficient that demand should persistently fall short of expectations and grow more slowly than capacity.[10]

That demand for electricity has lagged behind projections made by the utilities is a matter of record. According to Lovins's analysis, the greater the lead time and the more expansion of large power plants, the more utility's cash flow is encumbered, and at a considerable risk if demand is not as high as expected. He states the following:

> A higher growth rate should also increase the instability. How soon the cash flow starts to deteriorate should depend on the margin by which price elasticity is underestimated and on the safety margin built into the initial cash flow; utilities with low coverage ratio, high debt-to-equity ratio, low liquidity or high reliance on non-cash income has less room for maneuver. Conversely, utility managers alert

and flexible enough to respond to early warning signals by
drastically cutting back on construction (not merely defer-
ring it) may be able to stay out of trouble, or at least to
defer it. But the reflex response of most managers is to
suppress those signals by seeking higher subsidies, fast
depreciation, higher investment tax credit, and the like.
This is exactly the wrong thing to do; it makes the utility
crash harder (and somewhat later), and it thereby greatly
increases the risk to investors. Subsidy inflates demand
beyond the level whose revenues [the revenues of the
utility] can ultimately amortize the supply investment.
For a regulated utility which is required to provide adequate supplies and whose regulators are required to provide adequate return for it to do so, investment incentives
are also superfluous and can only encourage overbuilding
and overreliance on debt.[11]

Although subsidies such as investment tax credits and accelerated
depreciation help the utilities, they also distort the real financial
problems that accompany overbuilding of power plants. The fact
that these utilities are regulated monopolies further distorts the
real risk of overexpansion, since ultimately there will always be
ratepayers to protect them, at least in part, from the errors of excessive capitalization. Utilities are not subject to the market forces
that quickly cause other kinds of businesses to pay the price for
overbuilding and for production of supply, which substantially exceeds demand. Compared with the risk to utilities of underbuilding
base-load capacity, according to Lovins's calculations, "The calculated financial risk to utilities, shareholders, and ratepayers is
greater if baseload plants are overbuilt than if they are underbuilt
because the extra depreciation and return on excess baseload capacity will cost more than extra operation of short-lead-time peaking
and intermediate-load-factor plants (coal-fired or even gas turbines). This result is quite insensitive to the extra fuel costs of
such an operation or even of importing electricity" (p. 13).

To avert financial problems for the utilities, Lovins proposes
a transfer of capital away from overbuilding large power plants, including nuclear plants, and toward a program of conservation and re-
newable-sources loans. The loans would be at the internal interest
rate of the utility, that is, at a rate that automatically passes through
to the consumer whatever tax subsidies the utility benefits from.
Borrowers could use the loans in a number of ways and would repay
the loans through their utility bills. Thus, what the customer would

spend to improve energy efficiency would pay for itself in a lower bill from the utility, which, when coupled with monthly payments on the loan, would not exceed what the monthly bills had been before the conservation loan had occurred. After the loan is repaid, the full effect of the cost saving from greater energy efficiency would be enjoyed by the customer. The advantages to the utility of such a loan program compared with the current use of capital, according to Lovins, are that

> the alternative investment requires several times less capital than a new plant would;
>
> since the borrowers' investment typically takes days or months to build and a few years to pay back, compared to ten years to build and thirty to pay back for a power station, the utilities turn over their money much faster, improve their cash flow, and do more work per dollar of working capital, while providing a profitable home for any excess revenues from rate reform;
>
> by getting into a short-lead-time, fast-payback business at the margin, utilities remove the instability inherent in their cash flow and hence avoid overbuilding and eventual bankruptcy.[12]

Some providers of electricity are involved in conservation loan programs, in part for the advantages cited by Lovins. For example, S. David Freeman, chairman of the board of the Tennessee Valley Authority, explained the financial importance of conservation to the U.S. House of Representatives' Subcommittee on Energy Power, stating that "it is investment in energy conservation that yields the most benefits for every dollar of investment."[13] Freeman stated that utilities should play a leading role in conservation because

> they have a substantial interest and knowledge in the energy area already and have a positive financial stake in promoting conservation because it can eliminate the need for expensive new capacity. Utilities invest billions of dollars of capital each year. If this capital is not permitted to be used to fund conservation measures it will fund more expensive production and thus result in higher utility bills and more inflation.[14]

Later, he asserted that "the marginal cost of new generating capacity for this capital intensive industry has made energy conservation seem less of a threat and more of a promise of financial salvation."[15]

In keeping with this policy, the Tennessee Valley Authority (TVA) makes no-interest, home-insulation loans up to $2,000, the average loan totaling about $1,000, with the terms of repayment up to seven years. Its newly created Department of Energy Conservation and Rates will make conservation a long-term tool for reducing demand within TVA's large service area and reducing costly expansion of generating facilities. In addition to TVA, Pacific Power and Light Company began to offer no-interest conservation loans in 1978, as does Puget Sound Power and Light in the Seattle area, which also provides long grace periods on the loans because conservation saves them so much money.

Indeed, consensus is beginning to emerge within the electric-utility industry that conservation and alternative energy systems such as solar power make more business sense than do new, additional, massive centralized power-generating facilities. At a symposium on April 18 and 19, 1980, put on by the California Public Utilities Commission, two leaders of the electrical-utility industry voiced the same point of view. Charles Luce, chairman of the board of Consolidated Edison, New York, and Frederick Mielke, Jr., chairman of the board of Pacific Gas and Electric Company, both declared in strong terms that the new economic realities of the electrical power industry require utility companies to emphasize conservation and alternative energy sources much more than was the case in the past. Even more emphatic was another conference participant, Eugene Meyer, vice-president of the Kidder Peabody brokerage firm. Meyer's advice to utility management is to curtail new commitments for generating capacity, regardless of growth forecasts; pare current construction to the bone; and really encourage conservation efforts.[16]

Although there are some legal restrictions on how deeply utility companies may get into nontraditional parts of the energy-delivery system, and there are philosophical questions as to what constraints should be placed on them as they begin to do so, it is apparent that major companies in the electrical-utility industry see their role as having shifted away from simply adding new and larger generating plants to meet their customers' needs.

Even with the aggressive practice of conservation, there will remain a need for electrical power generation. Traditionally, the generation of electricity has come from a three-tiered, centralized system that includes large base-load power plants, intermediate-load facilities, and peak-load power plants. A base-load power plant is aimed at generating electricity as close to full time as possible and is considered to be the backbone of an efficient, reliable system, although in the case of nuclear power that reliability has been shown to be questionable. An intermediate-load plant generates electricity much of the time, but if demand is particularly low at any one time it may be temporarily shut down. A peak-load power plant, as its name suggests, is used only during periods of peak electrical demand and is more likely than the other two kinds of power plants to be run on higher-cost fuels such as oil.

In the past there was agreement that large base-load facilities were needed in order to generate electricity inexpensively, benefiting as they did from the economies of scale. Some observers maintain, however, that in the 1970s the economic benefits of massive centralized power generation disappeared, in part for the reasons discussed in Chapter 2. Carl E. Behrens points out that central power generation is afflicted with four kinds of uncertainty: uncertainty of future demand, construction-scheduling uncertainty, uncertainty of the cost estimates of new units and their alternatives, and uncertainty about the role of electricity in the evolving national energy picture.[17]

Decentralized smaller units afford advantages in dealing with these uncertainties. Clearly, smaller units afford greater flexibility in responding to demand and a possible changed role in meeting the nation's energy needs: fewer energy eggs would be put in the highly centralized generating-facility basket. As for scheduling the cost uncertainties, problems could persist if brand new technologies were used; but presumably old, known technologies would reap the benefits that accompany the construction of smaller, less costly facilities. Recent studies show that 500-megawatt facilities can be more economical than those of 1,000 megawatts. In view of the significant overbuilding of generating capacity that has occurred in the United States in the past, policies aimed at encouraging more numerous, smaller electrical-generating facilities would allow the greater flexibility needed to reduce the magnitude of that overbuilding.

Discussion of the appropriate scale for generating facilities is critically important to the issue of alternatives to nuclear and coal-powered electrical plants. If one believes that new 1,000-megawatt power plants are a necessity for the future, the impact of alternative fuel sources or generating systems is greatly reduced, because large plants tend to be fired either by coal or nuclear power. If, on the other hand, smaller power plants are encouraged, the possibilities of cost-competitive and environmentally superior fuel sources are greatly enhanced. The following alternative fuel or power sources are illustrative of the rich diversity of options available to meet the nation's electrical-generating needs. Regional differences will do much to determine the cost-effectiveness of each given fuel.

COGENERATION

Cogeneration has been described as an old idea whose time has come. It is defined as the simultaneous production of electricity and such useful thermal energy as hot water, steam, or hot gases. In the context of alternatives to nuclear power, cogeneration can best be described as the process of using industrial heat both for a given industrial purpose as well as to generate electricity. At the turn of the century, most industries requiring electricity generated their own power along with steam and whatever other process heat was needed. After 1920 the development of large central-station generation coupled with inexpensive fuels led to utilities taking over most power production. For the subsequent 50 years the cost of electricity delivered by utilities was such that it made economic sense for many companies not to generate their own electrical power. Tax laws further encouraged this trend away from cogeneration. However, the extremely high capital costs and fuel costs of central-power generation that became evident in the 1970s have helped create another sea change in the economics of cogeneration, such that it promises to become a major contributor to the nation's electrical power supply, of which now only 4 percent is derived from cogeneration.

The potential of cogeneration is indeed remarkable. A study by Richard Carlson, David Freedman, and Robert Scott found that the seven largest energy-using industries, including steel, oil refining, chemicals, and paper, could install cogenerating equipment by 1985 that would produce 46 gigawatts (or 46,000 megawatts) under current economic conditions.[18] The same study also suggested that a total of 167,000 megawatts could be produced by the seven industries if certain additional economic changes were made. Studies by Dow Chemical Company and Thermo Electron Corporation have also been strongly encouraging. Thermo Electron's study showed that by 1985 as much as 135,000 megawatts of power could be produced economically with gas turbines as a by-product of process steam generation at industrial sites. A U.S. Department of Energy study indicated that cogeneration could account for from 33,000 to 115,000 megawatt capacity under different scenarios by 1990.

How much cogeneration's potential is realized depends in large part on the extent to which certain institutional barriers and attitudes can be torn down or surmounted. Those barriers include the following:

1. High rates for stand-by power charged by utilities to industrial cogenerators,
2. Low prices paid for excess power from a cogeneration facility selling power to a local utility,
3. Significant investment costs required by a firm to become capable of cogeneration,
4. Fear of additional regulation by leaders in industries that could cogenerate,
5. Attitudes by leaders of potential cogenerating companies that electrical power generation is not their business.

The U.S. Congress attempted to deal with some of these problems in 1978 when it passed landmark energy legislation called the Public Utility Regulatory Policy Act (PURPA). PURPA represents a major commitment to cogeneration on the part of the federal government, the final success of which will depend on activities by the states. Among that law's most important provisions are the following:

1. The Federal Energy Regulatory Commission (FERC) provides guidelines for state regulatory commissions requiring that power be given to qualifying cogenerators at nondiscriminatory prices that are "just and reasonable." Vigorous implementation of nondiscriminatory pricing rules by public utility commissions will ensure fair prices to cogenerators when they need to buy electricity from utilities.
2. Utilities are required to purchase power from qualifying cogenerators at rates that are based on "avoided costs" to the utility. This may be the most important provision of PURPA. What it means is that both fixed and operating costs that a utility can avoid by getting its power from cogenerators serve as the basis for determining the price a cogenerator will get for its power when it sells it to a utility. By using a utility's avoided "incremental" cost, rather than the average system cost, the cogenerator should be assured an adequately high price for its electricity, in turn making cogeneration economically attractive.

3. PURPA also exempts qualifying facilities from much of the regulation required of conventional electrical utilities. As long as less than 50 percent of the cogenerator is not owned by an electrical utility, there is not a size limit, under guidlines published in the Federal Register of July 2, 1979. Moreover, a cogenerator would not be burdened by any of the rate-regulation requirements that electrical utilities must experience.

In addition to PURPA, the Energy Tax Act of 1978 adds tax credits to the existing business-investment tax credit to provide further incentive for investment in cogeneration. As well, the Windfall Profits Tax of 1980 provides a 10 percent business energy-investment credit (in addition to the regular 10 percent investment tax credit) on qualifying cogeneration facilities.

If the message from the Congress on cogeneration is clear and enlightened, the response on the part of the states is still uncertain. In deciding the basis for calculating a utility's avoided cost, a difficult job to say the least, the respective public utility commissions will do much to determine the true effect of PURPA. In what ways the other provisions of the federal law are honored also will do much to determine the future of cogeneration. If utility companies are pushed by regulatory commissions or others in state government to cooperate with and encourage cogeneration, the contribution of this old technology could even exceed the optimistic predictions already mentioned. That it is in the interest of the citizens of the several states to do this is clear: cogeneration helps avoid power-plant siting problems; reduces intrusive new power lines, since this more decentralized system can reduce the need for long power lines; causes fewer air-pollution problems; provides shorter lead times in construction, thus reducing capital costs; and further diversifies fuel sources. Public utility commissioners throughout the United States should be held accountable for the success of this very hopeful technology, which in West Germany accounts for about 30 percent of that country's electrical power.

WIND POWER

In the 1920s there was widespread and growing use of windmills or wind generators, a phenomenon that persisted in rural areas until the Rural Electrification Administration made them uneconomical in the 1930s, causing them to fade in importance. Rapid cost increases of traditional generating systems, however, have caused a rebirth of interest in wind power, in turn triggering growth in the business of manufacturing wind generators. The price and generating capacity of wind generators vary considerably: the range of cost is between $3,000 and $20,000, and the generating capacity is from one to ten kilowatts for a residence and/or farm. A homemade system will be less expensive, of course. It is likely that as the number of wind generators grows, few of them will serve as the sole source of electrical power for their owners, and additional costs will be incurred to pay for the necessary converter required to make the electrical system compatible with that of a utility. To help encourage the use of wind generators, the federal government has a tax credit of 40 percent on the first $10,000 of home wind-generator equipment and 25 percent for businesses investing in wind-generating equipment, in addition to which many states have additional tax credits or available rebates. Still, financing remains an important impediment to the swift development of this inexhaustible fuel source, and additional federal incentives analogous to those provided nuclear power, possibly in the form of loans, could do much to hasten the full use of this technology.

Even lacking a strong governmental push, however, the wind-generation business has blossomed in recent years. Whereas a decade ago there was only a handful of manufacturers of home and remote-site windmills, there were 23 small businesses making these devices in the spring of 1980, according to Ben Wolff, director of the American Wind Energy Association. Wolff estimates that the number of firms may decline as some firms succeed in making wind power profitable. The future generating potential is enormous, according to another wind power advocate, Ned Coffin, chairman of the Enertech Corporation of Norwich, Vermont. Coffin says that with wind machines of ten kilowatts or less, there could be 93 million kilowatts in installed wind power sometime in the next century.[19]

How practical and inexpensive wind power is depends on one critical variable: how much wind there is at the generating site. If the average wind velocity is less than 10 miles per hour, a wind generator would have difficulty being economically competitive; but if that average is 12 to 14 mph, it will likely be competitive with what the electricity would cost from a new centralized generating facility. Obviously, there is strong regional variation in average wind speed, and even one mile-per-hour difference in average velocity is very significant, since the energy output of a wind generator varies according to the square of the wind speed. Under the right conditions, the cost per kilowatt-hour can be as low as from four to eight cents, well within the existing range of the cost of centralized power plants and far less than the cost of new nuclear facilities.

Because the cost of storage of electricity from a wind generator is so high, it is not now economical for most owners of these devices to attempt total electrical self-sufficiency. That raises the important issue of interconnectibility with the existing electrical-generating system. While the Public Utility Regulatory Policies Act of 1978 requires utility company cooperation in interconnection with such decentralized facilities as wind generators, how that law is allowed to work depends largely on the good faith of the utility in question. An interconnection must meet the utility's own standards of safety, for example, and that could possibly be used as a deterrent to full cooperation. Similarly, utilities must allow the decentralized generating facilities to sell back to the utility any excess power those facilities might generate. Setting a fair price for that power, however, can be a sticky and complicated issue. As wind generators catch on, it is hoped that the early cooperation displayed by many utilities will persist.

HYDROELECTRIC POWER

A 1977 United States Army Corps of Engineers study indicated that the nation could double its existing hydroelectric generating capacity without conflicting with alternative water and land uses if existing dams were made capable of generating electricity. Most such new dams will be small-scale compared with such well-known facilities as the Grand Coulee Dam, but taken together, they can make a significant contribution to meeting electrical power needs. In New England, for example, a study by the New England River Basins Commission indicated that of 11,000 existing dams in that region, 1,950 could make a practical contribution to the region's power needs. The combined output of those dams could generate an additional 1,000 megawatts of added capacity, equivalent to nearly 7 percent of the 1978 New England winter peak-load demand and about 14 percent of the added electrical power capacity needed in the ten-year period of 1979-89.

The New England study indicated that some economic incentives are needed to help develop the full potential of hydroelectric power. Only 15 percent of the electrical-generating potential could be met under current conditions, but according to the report, if interest rates could be reduced to 3 percent and the return approached 67 mills per kilowatt hour (equivalent to the current New England power value for oil-fired intermediate generation), then 80 percent of the hydroelectrical power generation could be realized. If hydroelectrical power were allowed to benefit from federal subsidies on even a fraction of the scale enjoyed by nuclear power, a region such as New England would be able to move expeditiously toward full realization of its hydroelectric potential.

An encouraging development is a measure that was tacked onto the oil Windfall Profits Tax bill. It allows an 11 percent tax credit on projects smaller than 25 megawatts at existing dams, thus giving hydroelectric developers the same credits awarded other renewable-energy developers. Equally encouraging is the apparent shift in attitude in the financial investment community vis-à-vis hydroelectric power. For example, on March 27, 1980, Marine Midland Bank of New York announced its intention to invest $50,000 for development of a 1.5 megawatt plant in Wappingers Falls, New York. The dam site is one of the 26 targeted by New York State for electricity generation. Another indicator that shows the hastening interest in hydroelectric power was reported in the March 28, 1980, <u>Christian Science Monitor</u>. That paper reported that in the preceding five months the Federal Energy Regulatory Commission had received an increase in permit applications for development of dam sites from 78 to 161, and license applications had more than doubled over the previous year.[20]

The 1978 National Energy Act has triggered a new kind of "gold rush," according to John McPhee.[21] Dilapidated old dams have taken on a new allure to a growing group of entrepreneurs who want to rehabilitate them and make a tidy and sustained profit. McPhee cited two brothers from upstate New York, John and Jim Dowd, who educated themselves on how to reconstitute a run-down dam site and, with no state or federal loans, developed their plans to the point of obtaining solid bank financing to complete their restoration. When offered $250,000 for the site (their estimate of the total cost of reconstruction), the Dowd brothers turned it down flat, thus refusing an offer that would have constituted a 25 million percent return on their investment. The corporation that they turned down was one of a growing number set up after the 1978 legislation. These small companies consist typically of lawyers, engineers, and financiers, and McPhee maintains that for them "obtaining a preliminary permit from the Federal Energy Regulatory Commission is the equivalent of staking a claim in the search for gold." State regulators have much to say about how lucrative this new industry will be: in New Hampshire, where the state set the price at eight cents per kilowatt-hour, the interest in this energy form is high. Once the needed capital investment is made, the owners of these facilities have a clean and inexhaustible source of power, which can only increase in value as the cost of nonrenewable fuel continues to escalate.

Several regions have considerable potential for hydroelectric-generating capacity, as will be seen shortly in the discussion of a strategy for a nonnuclear future.

BIOMASS

When processed properly, wood and refuse can be used to fuel intermediate-sized power plants.

Wood

In 1977 Burlington Electric Company in Burlington, Vermont, converted a five-to-six-megawatt coal-burning boiler to one that burns wood chips for just $25,000. According to that utility's annual report for the year ending June 30, 1978, the wood-burning unit produced power at a rate of two cents per kilowatt-hour; the cost of producing coal in the same unit had been 30 to 35 percent higher. The conversion was so successful that Burlington Electric converted a second boiler to wood and, with the overwhelming approval of local voters, in 1978 issued bonds to construct a 50-megawatt facility to be completed by 1983.

In addition to obviating the need to buy more expensively elsewhere, the Burlington Electric project helped upgrade Vermont's forests by weeding out and using low-quality wood that choked and clogged the growth of high-quality timber. Furthermore, wood burning produces fewer pollutants than does coal burning: the ash left after burning is 1 percent from wood, compared with 8 percent from coal. There is no net increase of CO_2 because trees—the fuel source—consume as much of it while they are growing as they release when they are burned, and wood has no CO_2 emission.

The Burlington example may be catching on. In Minnesota, Northern States Power intends to convert its intermediate-load, 25-megawatt Red Wing plant from coal to wood, using wood from sawmills in Wisconsin and Minnesota and even burning old railroad ties.

Refuse

Using processed garbage as fuel is not a new idea; there are 181 refuse-processing plants in Europe. This fuel potential is just beginning to be realized now in the United States. Refuse can be used to fuel intermediate-load power plants or supplement coal in base-load facilities, as is done in Milwaukee, where roughly 30,000 of that city's 250,000 homes get their power from garbage. Among other cities, Chicago, Illinois; Ames, Iowa; Madison, Wisconsin; Akron, Ohio; and the Eastman-Kodak complex in Rochester, New York, currently use refuse to help generate electrical power. Large facilities capable of processing 3,000 tons of garbage a day are planned for Dade County, Florida, and Detroit, Michigan.

Unlike other fuels, such as those used in nuclear power, "garbage power" does not require massive government subsidies. For example, the Connecticut Resource Recovery Authority built a garbage-processing plant by issuing $53 million worth of revenue bonds, not backed up by the full faith and credit of the state and to be paid through waste-disposal charges to the nine affected cities in the Bridgeport area and through the sale of the processed garbage to the local Bridgeport electrical utility. Priced on a Btu-equivalent-to-oil basis, the fuel will be as cheap or cheaper than what was previously used by the utility, and the 900 tons of fuel produced each day will displace 650,000 barrels of oil a year.

GEOTHERMAL ENERGY

Energy can be derived from heat in molten rock deep within the ground, which, possibly through seismic activity, makes its way close enough to the surface to be used to generate electricity. Geothermal energy is used in Italy, Japan, Mexico, the USSR, and New Zealand to generate electricity, and there is one power plant in the United States that uses this energy source. At the Geysers in Sonoma County, California, Pacific Gas and Electric Company runs a 665-megawatt facility using 13 geothermal units, generating power that is 25 percent less costly than nuclear power. While there are still some technical problems that need to be solved—a way to remove corroding mineral impurities from the geothermal brines must be found, as well as a way to dispose of the liquid waste—stepped-up research and assistance by the government could make geothermal energy more attractive soon. The generating potential is considerable, according to Fred L. Hartly, the head of Union Oil Company, a firm that is actively exploring possible geothermal-generating sites in southern California. According to John Berger, author of Nuclear Power: The Unviable Option, in an open letter to employees and shareholders of November 3, 1975, Hartley said that at least 20,000 megawatts of electrical power could be generated from geothermal wells by 1985.[22] Depending on the assessment of the full environmental impact of geothermal power, this is an energy source with considerable potential.

PHOTOVOLTAIC ELECTRICITY

Through the use of the earth's most abundant solid element, silicon, it is now possible to convert sunlight directly into electricity. A photovoltaic cell is typically made up of two thin layers of silicon, one "contaminated" with boron and the other with phosphorous. When energy from the sun, in the form of photons, hits the cell, some of the electrons in the cell are displaced and establish an electrical current along the thin layer of aluminum that connects the two layers. Space ships and distant weather stations have already used photovoltaic cells to generate electricity.

The trouble with photovoltaic cells is that at this time they create a very expensive electricity. Although photovoltaic electricity is derived from the same semiconductor technology that has created inexpensive transistors, it will have to become ten times less costly to be competitive with conventional electrical sources. Reductions in cost have been encouraging, however. Photovoltaic cells are a hundred times less expensive now than the original satellite cells.[23] The main reason the cost remains so high is the cost of the appropriate form of silicon needed in a photovoltaic cell. There are encouraging words within the electronics industry, however, mainly relating to development at one of the nation's leading electronics companies, Texas Instruments.

For business reasons, much of the information relating to Texas Instruments's breakthrough in photovoltaic electricity is wrapped in secrecy. However, Benjamin Rosen has pieced together some of the essentials of Texas Instruments's breakthrough. The two key problems of high manufacturing costs and storage of electricity appear to be near solution, according to Rosen. Rather than using wafers, Texas Instruments has apparently come up with a way to manufacture spherical cells, which consume less energy to produce than do wafers and are more efficient and more reliable. As well, the company appears to have found a workable storage system based on stored hydrogen, which can, on demand, reproduce in reverse the process that turns sunlight into electricity.[24] The storage problem has been one of the critical issues in making photovoltaic electricity practicable.

According to Rosen, what this all could mean to the potential residential customer is a system of solar cells covering 2,000 square feet (45' × 45') on a roof, yielding up to 20,000 watts at a cost competitive with today's five to ten cents per kilowatt hour. It would require a $10,000 capital investment in 1990, a price that will seem modest compared with the cost of homes then. If the cost of central-station power generation continues to soar, electricity at the price Rosen describes will appear most attractive.

Within the next few years Texas Instruments will decide whether to make a major production commitment to this technology. The company has received $14 million from the U.S. Department of Energy (DOE) to help it research and develop photovoltaic electricity for the next four years. Compared with the billions of dollars that have been spent on nuclear power by the federal government, however, some observers are critical of the low priority and scant attention given to the photovoltaic alternative by the Carter administration and its predecessors.

Barry Commoner is one such person. Commoner, a presidential candidate and a professor of environmental sciences at Washington University in St. Louis, believes photovoltaic electricity could develop much faster with greater support from the federal government. Much of his argument rests on the idea that through its purchasing policies the federal government could do much to stimulate the development and production of photovoltaic electricity and reduce its cost. In his book The Politics of Energy, Commoner says that in 1978 the Federal Energy Administration (FEA) recommended a $400 million purchasing commitment on the part of the federal government, but the administration only recommended $98 million.[25] The smaller the federal government's commitment to photovoltaic electricity, the longer it will take to reduce its cost per kilowatt. While the Carter administration supported helping in the research and development of this technology, it refused to make the kind of commitment to a purchasing program on the scale recommended by the FEA and Commoner. Although the funding level had risen to $147 million for the fiscal year 1980 in the DOE budget, that figure was cut to $140 million for 1981.

Photovoltaic electricity presents a potentially serious threat to any utility that is solidly committed to centralized power generation. Photovoltaic cells will offer considerable opportunity for consumer control and choice, since more than one company will eventually get into the business.

> The cells could be used on site, like solar heating. But photovoltaics as a generator of electricity would have greater effect than solar heating on the role of utilities, for if on-site power generation by photovoltaics became widespread, the utility industry would have to redefine its business, emphasizing power distribution instead of power generation: utilities would have to purchase power from on-site installations, mark it up, and distribute it. Unless low-cost means of electrical power storage were developed, the utilities would still supply conventional backup power. A structural change of this sort would require imaginative solutions on the part of regulatory commissions, and would require the support of the utilities themselves, which would by no means be certain.[26]

The study goes on to say that the utilities would likely favor photovoltaic electricity only if it could be used to generate power from a centralized facility as is done now with coal and nuclear power. Yet this technology's greatest asset is the extent to which it allows decentralized generating capacity and permits competition for the market, thus eventually pushing costs down.

CENTRALIZED SOLAR POWER

Although more economically suited to decentralized use, solar power in the form of small, centralized thermal-energy units appears technologically feasible. Solar collectors would convert the sun's energy by heating water, and the heat energy from the water would be converted into electricity by turbine generators similar to conventional systems. The U.S. Department of Energy, in cooperation with the local utility and the electrical power industry, is building an experimental ten-megawatt facility in Barstow, California. The trouble with this means of generating electrical power is that massive amounts of land are needed for the large solar collectors, and currently the cost per kilowatt hour of this kind of power is not even close to being competitive with other means of generating electricity. Unfortunately, if public attention is focused on this unwieldy technology, it will reduce the credibility of the realistic and attainable alternatives to nuclear power that already make sense today.

CENTRALIZED WIND POWER

Using wind power to generate electricity on the centralized basis is also largely in the demonstration stage. While the DOE and others are researching its practicality, the cost per kilowatt-hour is still too high to permit swift implementation of this option. Again, however, were wind energy to benefit from the same kind of sustained subsidies enjoyed by nuclear power, the picture could change dramatically. The potential energy from wind is enormous. As a civil engineer and professor at the University of Massachusetts, William Heronemus estimated in 1972 that a large offshore wind-power system could provide electricity needed for the six New England states more cheaply than could nuclear power.[27] According to John Berger,

> Heronemus estimates that the total cost of installing such a wind system, complete with hydrogen energy storage, would be about $22.4 billion. Hydrogen can readily be produced from seawater by electrolysis, the passage of an electric current through the water, causing the dissociation of water into oxygen and hydrogen. The product gases can be collected at opposite terminals of an apparatus. Direct current generated by the wind would be used to perform the electrolysis, and the resulting hydrogen and oxygen would be pumped ashore where they could be recombined as needed in a fuel cell to produce electricity.[28]

Simple technology would be required, compared with that of nuclear reactors, and Heronemus stated his belief that with proper impetus, such a system could be operating within 24 months of the program's inception. Southern California Edison, Bonneville Power in Oregon, and Northern States Power in Minnesota are some utilities actively exploring the potential of centralized wind power.

POWER FROM THE SEA

Tidal power can be harnessed by building a dam across the mouth of a bay to tap into the power generated by the rising and lowering tides. However, few locations are well suited to using tidal power: according to John J. Berger, only one site in the United States is even a possibility,[29] and the only such facility working in the world operates in an estuary at La Rance, France. More promising, according to Berger and others, is ocean thermal energy conversion (OTEC), which takes advantage of temperature differences of water in tropical oceans to power a heat engine. The heat differential causes a liquid with a low boiling point (such as Freon) to vaporize, and the force of the vapor drives a turbine, which provides the energy to produce electricity. The technology of OTEC has existed for half a century. In 1929 George Claude, the French scientist, demonstrated a 22-kilowatt ocean thermal plant at the Bay of Matanzas in Cuba. The southeastern coast of the United States is suitable for development of ocean thermal power, and once it is fully developed, according to Berger, it would supply vast quantities of energy, possibly exceeding all such needs of the nation. While not currently competitive with traditional generating systems from the standpoint of cost, accelerated development of this technology, combined with the cost escalation of uranium and fossil fuels, could make ocean thermal power an attractive option soon. Moreover, its application will come in a region, the Southeast, which needs alternatives to the expensive fossil fuels upon which it currently relies too heavily.

A STRATEGY TO PHASE OUT NUCLEAR POWER

Two points must be made to put the issue of phasing out nuclear power into proper perspective. First, nuclear power accounts for only about 4 percent of the nation's total energy supply now, and only 13 percent of its electrical-generating capacity. Thus, a planned phasing out of nuclear power would not require a massive shift in the nation's energy supply. Second, there is currently a large excess of electrical-generating capacity in the United States. According to Steve Nadis, the Federal Power Commission recommended a reserve margin of 20 percent, and in 1978 generating capacity exceeded demand by 33 percent.[30] Nadis maintains that if all existing nuclear-generating capacity were brought off line, the remaining capacity would still provide a reserve margin of 21 percent nationally. If this occurred, in the most heavily nuclear regions margins would be at 6 or 7 percent. The point is that significant excess reserves make a planned phasing out of nuclear power even more practicable than would otherwise be the case.

Richard Carlson, David Freedman, and Robert Scott, all of Washington University in St. Louis, spell out both how nuclear power could be radically curtailed at once and how estimated growth in demand for electricity could be met in the future through nonnuclear means. Their strategy for the immediate closing of 64 of the existing 72 nuclear facilities is well reasoned and documented, if unlikely to be adopted. They maintain that by using fossil-fueled power plants to their full capacity and maximizing the potential for interregional power transfers (see Figure 5.1), all but eight nuclear power plants could be closed down immediately.[31] Part of the reason this scenario appears unrealistic is that it would require an 11 percent increase in the consumption of coal and a 37 percent increase in oil consumption, causing the total projected national electrical bill to rise from $92.7 billion to $97.5 billion. Even if the United States were to accept a 5.2 percent increase in the cost of electricity, it is unlikely that such a major increase in the use of oil for generating electricity would be accepted. Furthermore, such additional reliance on fossil fuel would result in correspondingly greater violence to the environment. Still, this part of their analysis does a service by emphasizing the potential for interregional power transfers and affording a systemwide approach. Their most valuable contribution, however, relates to how to meet future electrical demand without the heavy reliance on nuclear power that the industry maintains is needed.

The National Electric Reliability Council (NERC) has said that by 1987 twice as much of the nation's electrical power will have to come from nuclear power than does now—or 27 percent. That means that the 94 additional plants that now have construction permits would actually be built. Carlson, Freedman, and Scott see some serious flaws in NERC's estimates, perhaps the most serious of which is the demand estimates used by the electrical industry. Consider the following facts: in 1974 the utilities predicted an 8.8 percent increase in peak demand, and the actual increase was 1.6 percent. For 1978 the industry estimated a 6.2 percent increase but experienced only a 2.7 percent increase. During the first quarter of 1980, consumption actually dropped 1.4 percent. Rather than relying on data from a 20-year period, some of which time the cost of electricity actually dropped, the Washington University researchers decided to use only the period since 1973, during which time electricity prices increased, believing that this would provide a more accurate picture of future demand for electricity. Their estimates (in percent) and those of NERC are compared in the following table:

FIGURE 5.1

Nuclear-Power-Plant Locations and Interregional Power-Transfer Capabilities

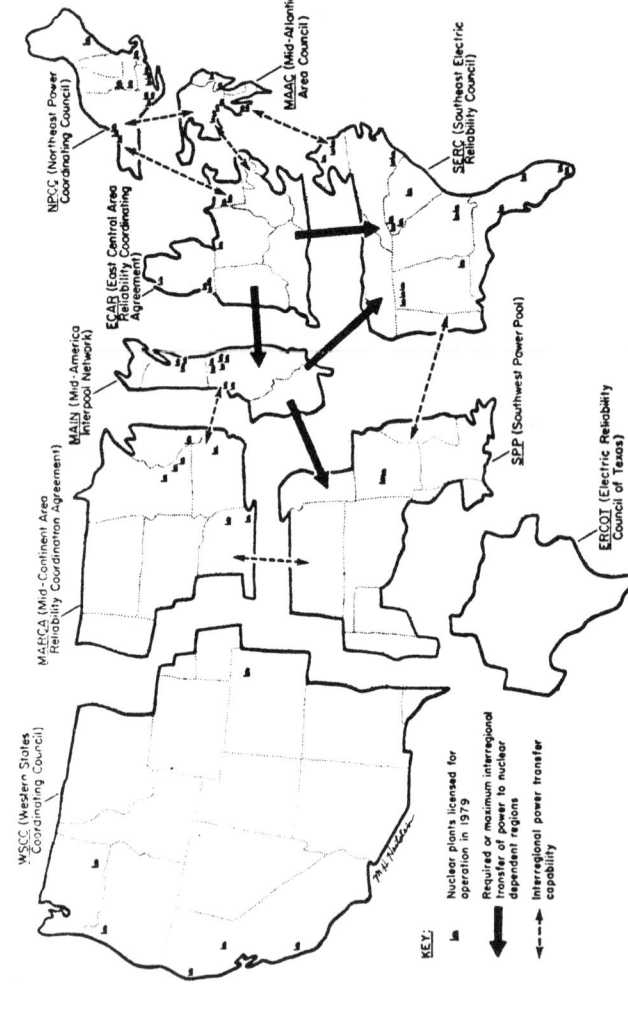

Source: Reprinted, by permission, from Richard Carlson, David Freedman, and Robert Scott, "A Strategy for a Non-Nuclear Future," Environment, July-August 1979, p. 8. Map by Marcy H. Harstein.

	Peak	Total
NERC	5.3	5.4
Carlson, Freedman, and Scott	3.5	3.6

If, as appears likely, the three researchers are closer to being correct, that would cut in half the projected additions planned by NERC. (In recent years demand has actually risen at only 2.5 percent per year.) However, the question of how to meet the remaining shortfall remains. Ignoring for the moment the tremendous potential that conservation, wind, and other alternatives provide, let us see how the researchers suggest that the job be done.

They begin by breaking down the United States into nine regions (Figure 5.1). In six years, Carlson, Freedman, and Scott estimated the required additions to existing generating capacity as indicated in column 1 of Table 5.1. Column 2 represents estimates made by NERC of expected nonnuclear additions to generating capacity; it shows that in four of the nine regions there would be excess growth in capacity even without stepped-up use of hydroelectric power and cogeneration. The third, fourth, and fifth columns show the potential for various nonnuclear means of generating electricity. The hydroelectric estimates come from research done by the United States Army Corps of Engineers and assume the use of existing dams for generating electricity because of the savings involved in using existing facilities. For example, the researchers indicate that, recently, new hydroelectric plants have cost an average of $616 per kilowatt-hour, but only $274 of that went to the structure, improvements, and equipment needed to generate electrical power.

By 1985 industrial cogeneration could account for 46 megawatts, according to a study of the seven largest energy-using industries cited by the researchers; this is assuming that no special economic incentives would be given to the companies involved other than the inherent incentive of more efficient—and hence, cost-saving—use of their energy. Moreover, the researchers add the following:

> Fortuitously, the largest industrial potentials for cogeneration are generally in those regions, such as MAIN, NPCC, and SERV, which are most reliant upon nuclear power. For example, cogeneration in the steel industry can contribute to eliminating the need for future additions of nuclear power in the Chicago-Gary area, while paper and chemical plant cogeneration can eliminate the need for new nuclear plants in the South.[32]

TABLE 5.1

Potential for Meeting 1987 Projected Demand without Nuclear Power
(in gigawatts)

	Required Additions to 1979 Non-nuclear Capacity to Meet 1987 Peak Demand*	NERC Projected Nonnuclear Net Additions, 1980–87	Additional Generation Potential, 1980–87			
			Hydro	Industrial Cogeneration	Residential and Commercial Cogeneration	Total
ECAR	9.4	31.4	3	7	11	52.4
ERCOT	11.4	6.2	0	7	3	16.2
MAAC	1.6	3.8	1	5	7	16.8
MAIN	10.5	9.5	1	4	8	22.5
MARCA	7.3	7.7	4	1	4	16.7
NPCC	2.6	2.7	6	3	10	21.7
SERC	44.8	27.1	7	9	10	53.1
SPP	38.1	23.6	4	6	5	38.6
WSCC	25.6	25.4	15	4	10	54.4
United States	151.5	137.4	41	46	68	292.4

*Projected peak demand for 1987, plus required reserve margin, minus 1979 nonnuclear capacity. ERCOT, MAIN, MARCA, and SPP have 15 percent required reserve margins; all other regions have 20 percent required reserve margins.

In addition to the greater energy efficiency that industrial cogeneration provides, they point out the "cost differential over simple boilers of $200 to $760 per kilowatt, compared with nuclear plant costs of $650 to $935 per kilowatt (both in 1976 dollars) for plants coming on-line at present" (p. 14).

Combining hydroelectric potential and industrial-cogeneration potential, there remains only one region, SERV, that would appear to have a shortfall, and that is only 1.7 gigawatts. Carlson, Freedman, and Scott advocate the use of a cogeneration device manufactured by the Fiat Company in Italy, which is based on a four-cylinder automobile engine that uses natural gas to cogenerate heat and electricity and could be used in residences or commercial establishments. Whether this device, called TOTEM, or simply conservation were used, it is likely that the southeastern United States could find a nonnuclear alternative to meet their remaining electricity needs.

Three main points can be made about the analysis done by the three researchers at Washington University.

1. Their demand estimates are much more closely related to recent actual experience than are those of the National Electric Reliability Council. The demand forecasts of NERC would appear to consign the country to even more overbuilding of generating facilities, to the economic and social disadvantage of its citizens.

2. Whether one accepts the strategy proposed by the researchers or some other strategy perhaps more closely tied to conservation, it is apparent that alternative means of meeting the electrical needs of the United States could become a reality if there were adequate political commitment and planning done to ensure that.

3. All the alternatives presented by Carlson, Freedman, and Scott are immediately available, as is conservation (if that were to become the cornerstone for a nonnuclear future). While they state that photovoltaic electricity might become available after 1987, their nonnuclear plan does not require it for the period from the present to 1987. Stepped-up use of hydroelectric power and increased industrial cogeneration can be developed in relatively short order. The question changes from one of whether a nonnuclear future is realistic to one of whether there is sufficient vision and will to ensure that it will happen. If it wanted to, the federal government could certainly devise an effective nonnuclear strategy for the future, possibly even one superior to that proposed by Carlson, Freedman, and Scott.

CONCLUSIONS AND COMMENTS

Reading literature on alternatives to nuclear power generation led to the following conclusions.

1. There are sufficient alternative technologies to begin phasing out nuclear power.
2. The most important technology for a nonnuclear future is the least glamorous: conservation. Los Angeles and Seattle showed it can be done.
3. There is tremendous diversity between regions with regard to the most appropriate alternatives to nuclear power; for example, hydroelectric and wood power can make a difference in New England, and geothermal, and possibly solar, power can become very important in the Southwest.
4. The use of wind, biomass, and cogeneration appear to have application throughout the United States, depending on local conditions.
5. Changes in the system of pricing electricity could help reduce the growth in demand for electricity and prevent further overbuilding of electrical capacity.
6. The move to decentralized electrical power is likely to continue as traditional systems become more expensive. Wind generators and photovoltaic cells may radically change the nature of the electrical power system in this country and, eventually, may change the very role of utility companies and other centralized providers of electricity.
7. With the right kind of leadership, a nonnuclear future is possible with a minimal increase in the reliance on coal. If necessary, however, coal is a preferable transitional fuel for a future that is to be more reliant on renewable resources because it does not have the long-term waste-disposal problems that plague the use of nuclear power.
8. A national strategy for a nonnuclear future is possible, as was shown by the three researchers from Washington University.

During the debate on my bill in late March 1980, I told my colleagues in the Minnesota House of Representatives of the various alternatives to nuclear power within our state. They listened patiently. I stressed all that wind that blows in from North Dakota in the northwestern part of the state and the wind around Lake Superior, as well as the potential we had to use hydroelectric power, cogeneration, and biomass here. Those arguments and those about the new economics of nuclear power, the problems of waste disposal, and nuclear safety were not enough. The bill failed, 58 to 73. I had learned a lot about nuclear power and felt good about the attempt I had made to pass what I still consider to be a very moderate bill. I simply could not get the votes.

Even with the nuclear-waste legislation dead, the research I had done made the issue very much alive for me. I had traveled a fascinating path of inquiry and concluded that nuclear power had become far more expensive than I had ever imagined, distressingly dangerous, and a far-from-necessary technical alternative for the nation to meet its electrical power needs. One question remained: If the options are available, why is there so little push from the major institutions of power in our society to achieve a phasing out of this dangerous energy source? Why does the public put up with it?

NOTES

1. Amory B. Lovins, "Energy Strategy: The Road Not Taken," Foreign Affairs 55 (October 1976): 78.
2. Ibid., p. 79.
3. Charles J. Cicchetti, Edward Berlin, and William J. Gillen, Perspective on Power: A Study of the Regulation and Pricing of Electric Power (Cambridge, Mass.: Ballinger, 1974), pp. 13-53.
4. Ibid., p. 38.
5. National Association of Regulatory Utility Commissioners, National Association of Regulatory Commissioners' Annual Report on Carrier and Utility Regulation (Washington, D.C.: NARUC, 1979), p. 617.
6. Marc H. Ross and Robert H. Williams, "The Potential for Fuel Conservation," Technology Review, February 1977.
7. Dean Abrahamson, "Environmental Costs of Electrical Power," Scientists Institute for Public Information, New York, 1970.
8. Amory B. Lovins, "Electric Utility Investments: Excelsior or Confetti?" (Remarks to the E. F. Hutton Fixed Income Research Conference on Public and Investor Owned Electric Utilities, March 8, 1979, New York).
9. Ibid.
10. Ibid., p. 10.
11. Ibid., p. 11.

12. Ibid., p. 16.

13. U.S., House of Representatives, Committee on Interstate and Foreign Affairs, Subcommittee on Energy and Power, <u>Utility Participation in Energy Conservation</u>, speech presented by S. David Freeman, October 12, 1979, Washington, D.C., p. 2.

14. Ibid., p. 3.

15. Ibid., p. 4.

16. Eugene Meyer, untitled speech to Symposium at Stanford University, Palo Alto, Calif., April 19, 1980, in <u>Energy Efficiency and the Utilities: New Directions</u> (San Francisco: California Public Utilities Commission, July 1980), p. 141. Meyer's advice to utility companies included the following seven points:

> 1) Regardless of what your load growth forecast indicates, make absolutely no new commitment for future generating installations unless you are absolutely certain: Regulation will employ a forward-looking test year throughout the construction period; regulation will allow full CWIP on base load plants in ratebase without offsetting AFUDC throughout the construction period; that needed regulation will approve normalized accounting for deferred taxes and the investment tax credit throughout the construction period; and you will be able to earn the market-determined cost of security capital throughout the construction period on an equity ratio high enough to maintain bond ratings of AA of 1A at the minimum. 2) pare current construction requirements to the bone by stretching out through slowdown and postponement of those projects which are already underway—regardless of the fact that those projects will thus be completed at higher ultimate cost. 3) do your best to cut nonfuel operating expenses across the board by at least 10% even though, in your judgment, reliability may be adversely affected. 4) work together with regulation to develop plans for rationing power resources during needle peaks and outages, forced or scheduled. 5) encourage conservation with all your heart, where such action will not increase your capital requirements. 6) under no circumstances be so foolish as to presume you can finance generating equipment, however large or small, for others when you already cannot do so for yourselves. 7) review this program annually to see whether it needs to be tightened up some.

17. U.S., House of Representatives, Committee on Interstate and Foreign Affairs, Subcommittee on Energy and Power, "Decentralizing Electric Power Generation: Technologies, Goals and Consequences," speech presented by Carl E. Behrens, in Centralized vs Decentralized Energy Systems: Diverging or Parallel Roads? Committee Print no 96-IFC 17, Washington, D.C., May 1979, pp. 16-17.

18. Richard Carlson, David Freedman, and Robert Scott, "Strategy for a Non-Nuclear Future," Environment 21 (July-August 1979): 12.

19. Frank Farwell, "New Energy: A Burgeoning Business in Windmills," New York Times, April 27, 1980, p. F-3.

20. Scott Armstrong, "Little Hydro Dams Make a Big Splash," Christian Science Monitor, March 28, 1980, p. 2.

21. John McPhee, "Minihydro," New Yorker, February 23, 1981, p. 77.

22. John J. Berger, Nuclear Power: The Unviable Option (Palo Alto, Calif.: Ramparts Press, 1976), p. 293.

23. Robert Stobaugh and Daniel Yergin, Energy Future (New York: Random House, 1979), p. 209.

24. As Rosen describes it,

> Thus, sunlight is converted to electricity which, through electrolysis, generates hydrogen gas. The hydrogen, in turn, is converted into a hydride which is burned in the fuel cell—as required, when required—to create electricity. The system is a closed one capable of continuous operations. Because the reactions in the tube and in the fuel cell can be made to approach thermodynamic reversibility very closely, they thus provide a highly efficient storage and generation system. [P. 7]

Benjamin Rosen, Electronics Letter (Morgan Stanley Co., New York), May 15, 1979.

25. Barry Commoner, The Politics of Energy (New York: Alfred A. Knopf, 1979), p. 38.

26. Stobaugh and Yergin, Energy Future, p. 209.

27. William Heronemus, "Pollution-Free Energy from Off-Shore Winds," Marine Technology Society, Washington, D.C., September 1972.

28. Berger, Nuclear Power, pp. 302-3.

29. Ibid., p. 311.

30. Steve Nadis, "Time for Reassessment," Bulletin of Atomic Scientists, February 1980, pp. 37-44.
31. Carlson, Freedman, and Scott, "Strategy," p. 10.
32. Ibid., p. 14.

REFERENCES

Articles

Armstrong, Scott. "Little Hydro Dams Make a Big Splash." Christian Science Monitor, March 28, 1980, p. 2.

"Back to Forest." U.S. News and World Report, April 23, 1979.

Beck, Melinda. "Energy: Start Small." Newsweek, October 8, 1979, p. 34.

Bryan, Richard W. "Getting Serious about Wood Fuel: Public Utility Begins Conversion." Forest Industries 106 (May 1979): 31-33.

Carlson, Richard, David Freedman, and Robert Scott. "Strategy for a Non-Nuclear Future." Environment 21 (July-August 1979): 6-15, 37.

Carter, Luther. "Academy Energy Report Stresses Conservation." Science 207 (January 25, 1980): 385-86.

Farwell, Frank. "New Energy: A Bourgeoning Business in Windmills." New York Times, April 27, 1980, p. F-3.

Lanouette, William J. "A Latter-Day David Out to Slay the Goliaths of Energy." National Journal, October 1, 1977, pp. 1532-33.

Lovins, Amory B. "Energy Strategy: The Road Not Taken." Foreign Affairs, vol. 55, no. 1 (October 1976).

_____. "How to Finance the Energy Transition." Not Man Apart 8 (September-October 1978): 8.

_____. "The Nuclear Future: Wrong Questions, Wrong Answers." Science, October 1979, pp. 21-25.

McPhee, John. "Minihydro." New Yorker, February 23, 1981, p. 77.

Nadis, Steve. "Time for Reassessment." Bulletin of Atomic Scientists, February 1980, pp. 37-44.

Rosen, Benjamin. Electronics Letter (Morgan Stanley Co., New York), May 15, 1979.

Ross, Leonard. "Commoner's Assessment of Nuclear Energy." Esquire, June 19, 1979, p. 19.

Ross, Marc H., and Robert H. Williams. "The Potential for Fuel Conservation." Technology Review, February 1977, pp. 48-57.

Stevens, Mark. "Utilities Remain Committed to Nuclear Power." Christian Science Monitor, May 11, 1979.

"T.I. Breaks a Barrier in Solar Energy." Business Week, June 4, 1979, p. 64.

Reports

Abrahamson, Dean. "Environmental Costs of Electrical Power." Scientists Institute for Public Information, New York, 1970.

Burlington Electric Department. "Burlington Electric Department Annual Report." Burlington, Vt., June 30, 1978.

Dow Chemical Co. Dow Chemical Company Study, PB-243823. Washington, D.C.: National Science Foundation, 1975.

Klock, Thomas E., and Dennis J. Waskiewicz. "Potential for Hydropower Development at Existing Dams in New England: Physical and Economic Findings." New England River Basin Commission, September 1979.

Maycock, Paul D. Overview of the National Photovoltaics Program. U.S. Department of Energy Abstract, Washington, D.C., 1978.

National Association of Regulatory Utility Commissioners. National Association of Regulatory Commissioners' Annual Report on Carrier and Utility Regulation. Washington, D.C.: NARUC, 1979.

Tennessee Valley Authority. "Program Summary, Division of Energy Conservation and Rates." Knoxville, Tenn., October 1979.

Thermo Electron Corp. A Study of Inplant Electric Power Generation in the Chemical, Petroleum, Refinery and Paper and Pulp

Industries, PB-255659. Prepared for the Federal Energy Administration, Office of Energy Conservation, Waltham, Mass., June 1976.

U.S., Department of Energy. Special Budget Issue. Washington, D.C.: Energy Research Digest, January 29, 1980.

Speeches

Lovins, Amory. "Electric Utility Investments: Excelsior or Confetti?" Remarks to the E. F. Hutton Fixed Income Research Conference on Public and Investor Owned Electric Utilities, March 8, 1979, New York.

Meyer, Eugene. Untitled speech to Symposium at Stanford University, Palo Alto, Calif., April 19, 1980. In Energy Efficiency and the Utilities: New Directions. San Francisco: California Public Utilities Commission, July 1980.

U.S., House of Representatives, Committee on Interstate and Foreign Affairs, Subcommittee on Energy and Power. Decentralizing Electric Power Generation: Technologies, Goals and Consequences. Speech presented by Carl E. Behrens, in Centralized vs Decentralized Energy Systems: Diverging or Parallel Roads? Committee Print no. 96-IFC 17, Washington, D.C., May 1979.

——. Utility Participation in Energy Conservation. Speech presented by S. David Freeman, October 12, 1979, Washington, D.C.

Books

Berger, John J. Nuclear Power: The Unviable Option. Palo Alto, Calif.: Ramparts Press, 1976.

Cicchetti, Charles J., Edward Berlin, and William J. Gillen. Perspective on Power: A Study of the Regulation and Pricing of Electric Power. Cambridge, Mass.: Ballinger, 1974.

Commoner, Barry. *The Politics of Energy*. New York: Alfred A. Knopf, 1979.

Stobaugh, Robert, and Daniel Yergin. *Energy Future*. New York: Random House, 1979.

6

HIDDEN POWER

There are powerful roadblocks to a nonnuclear future; understanding what they are can help start the process of overcoming them. The roots of nuclear power explain much about the formidable coalition that continues to advocate this energy form even in the face of drastically rising costs, risk to the safety of the public, and quickly attainable alternatives. That coalition, made up of giant energy corporations, electrical power businesses, and the federal government, can be considered the heart of the nuclear power industry. It may seem unusual to consider government as part of an industry, but in this case it is appropriate and fair to do so.

It should not be forgotten that the federal government, not private business, came up with this technology. After using it to create mass destruction and end a war, the government sought ways to make nuclear power a force for social and economic good. One of the main missions of the Atomic Energy Commission (AEC) was to promote the use of nuclear power for generating electricity. At first the utilities were wary of this potentially dangerous way of generating power. The first few reactors were actually subsidized by manufacturers like General Electric as a way of developing a new product and a new market. To make nuclear power even more attractive, the federal government got into the insurance business, offering to insure any power plant up to $560 million in case of liability suits that might result from an accident.

Far from being arm's-length overseers of the development of a new energy form, the federal government has been an eager advocate of nuclear power from its inception, a fact that is as true today as it was 25 years ago. Strong institutional resistance to a nonnuclear future persists within both the Nuclear Regulatory Commission (NRC) and the Department of Energy (DOE), but there is some reason to be optimistic that such resistance will diminish during the next ten years. For example, many of the people who ushered in "Atoms for Peace" will be retiring soon. Still, there are potent bureaucratic reasons for the younger, middle-management employees of the NRC and DOE to at least profess a pronuclear point of view: their bosses by and large think it is a necessary and desirable technology, Three Mile Island or not.

A personal experience of mine helped clarify the relationship between government and business with regard to nuclear power. During the summer of 1979 I sent for information from the U.S. Department of Energy concerning capacity factors of nuclear power plants. It seemed that the government could provide independent information on this important measure of nuclear power's dependability. In response to my inquiry, the DOE sent me a report written and distributed by the Atomic Industrial Forum, the nuclear power trade organization. Even disregarding the self-serving nature of the data in the report, the very fact that a major federal department would be satisfied with simply passing along to the public industry-generated information of such importance is distressing. Far from being an independent source of information, it would appear that the Department of Energy is merely a conduit of information from the private sector to the public.

The fact that the federal government is an active partner in the nuclear power industry is of inestimable importance when one assesses the true power of that industry, since the private portion of the industry is not exactly made up of "mama and papa" small businesses. Companies such as Union Carbide and Exxon are involved in mining and milling uranium, Westinghouse and General Electric manufacture nuclear reactors, and many utility companies throughout the nation possess considerable economic and political clout. The resources available to that kind of coalition are formidable, and the ability to shape popular opinion on the issue—particularly in contrast to those advocating a nonnuclear future—is almost overwhelming. Government and business have both played an important part in selling the people of the United States on the idea that nuclear power is necessary for the future.

SELLING THE IDEA OF NUCLEAR POWER

The private sector of the nuclear power industry both sells the good news and suppresses the bad news about nuclear power. The advertisement depicted in Figure 6.1 is typical of the advertisements that heavily pronuclear utilities are running throughout the country. Even if one ignores the misleading nature of the text, the main point about an advertisement like this one is that only pronuclear elements are able to afford this level of expenditure. Organizations like the Sierra Club and the Union of Concerned Scientists do not have the tens of millions of dollars needed to communicate a contrary point of view, and as a result, the U.S. public is left with grossly incomplete or erroneous information upon which to make a judgment.

In addition to publishing positive information about nuclear power through advertisements, the private part of the electrical power industry is also engaged in a massive public relations campaign to improve the atmosphere in which the nuclear power issue is debated and discussed. Through the sponsorship of the Edison Electric Institute, a nationwide celebration of Thomas Edison and electricity took place in 1980. There is little doubt that much of the motivation behind this campaign was to regain public confidence and trust for those who deliver electricity to the people of the United States. The better a person feels about a utility company, the more acceptable its decisions on nuclear power will be to one who is uncommitted on that issue.

Keeping the bad news about nuclear power away from the public eye is another part of selling it to people. A classic example of this occurred at Exxon Corporation. As described by Dan Dorfman, Exxon did an in-house study of nuclear power that found it far less economical than had been thought. Richard Hellman, once an assistant to the chairman of the Federal Power Commission and now a professor of economics at the University of Rhode Island, was given a briefing by Exxon researchers. As Dorfman describes it,

> They concluded . . . that there was no competitive advantage to nuclear power, that the use of coal was at least as cheap or cheaper, and that a meaningful nuclear investment by Exxon was questionable until the problems (both in safety and economics) could be resolved.[1]

This study, undertaken at the request of top management of Exxon, never saw the light of day. When I asked a public relations official

FIGURE 6.1

Middle South Utilities Advertisement

Source: Reprinted, by permission, from Fortune, September 10, 1979.

at Exxon for a copy of the report, he refused my request and belittled the report, indicating that it had been done by the "coal people" at Exxon, obviously a biased group. Moreover, he said, "Coal is great, except you can't mine it and you can't burn it." Unfortunate though it may be, perhaps Exxon's reluctance to release its study is understandable; its uranium mine in Douglas, Wyoming, is reportedly one of the largest in the United States.

The governmental part of the nuclear power industry has played an important part in emphasizing the positive and playing down down the negative aspects of nuclear power. Its role as founder, subsidizer, and promoter of nuclear power has already been discussed. It would be surprising if energy leaders in the federal bureaucracy were to be skeptical of this technology, since it is largely their creation. Of those who have occupied responsible government positions, no one is a bigger supporter of nuclear power than James Schlesinger, former secretary of defense, former head of the Central Intelligence Agency, and former secretary of energy. In his final pronunciamento as head of the DOE, Schlesinger said, "Quite bluntly, unless we achieve the greater use of coal and nuclear power over the next decade, this society may just not make it." The power of opinion of persons like Schlesinger should not be discounted, ill-considered and erroneous though his remarks may have been. Even though he began and ended his job as energy secretary with a strong pronuclear bias, his views carry the weight of a cabinet member. People like Schlesinger not only have the power to be heard but also can do much to silence critics within the nuclear power industry and keep damaging information from being made available to the public. Two examples are illustrative.

The suppression of information damaging to the nuclear power industry is not a new phenomenon. In a very well-documented portion of <u>The Menace of Atomic Energy</u>, John Abbotts and Ralph Nader present strong evidence indicating that the Atomic Energy Commission (AEC) withheld publication of an update of a reactor accident study, called WASH-740.[2] The original study, done in 1957, indicated that an accident could kill 3,400 people, injure 43,000, and cause $7 billion in property damage. Those preparing the update came up with even worse figures: a potential 45,000 dead, 100,000 injured, and property damage of $17 billion. At the urging of the Atomic Industrial Forum, the AEC did not make the unhappy news public when it was ready in 1965. General acknowledgments were made that bigger reactors and more fuel might make an accident more dangerous, but the AEC did not see fit to release the unflattering figures until 1973. During that interval the AEC actually denied the existence of a the update, and it was apparently only the prospect of a law suit brought under the Freedom of Information Act that motivated the AEC to release 2,100 pages of working papers on the update—eight years after they had been available.

In 1964 Thomas Mancuso was commissioned by the Atomic Energy Commission to study the health effects on workers who labored in various nuclear facilities. Mancuso recounted a story of government interference in a lengthy and detailed research project that was well on the way to establishing a case that the effects of low-level radiation were worse than what had previously been thought.[3] That story is worth telling, since it reflects on how the federal government can deal with researchers who fail to come up with the "right" answer.

Mancuso compiled data on over 200,000 workers, stretching over a period of 30 years. Arduously collating the data, Mancuso looked at job history, amount of radiation suffered, cause of death, and other factors in his attempt to find a correlation, if any existed, between death rates and exposure to radiation. Because of the voluminous material and sometimes fragmentary information, he did not expect to reach any conclusions until the late 1970s. In 1974, however, he suddenly received pressure from Sidney Marks, then the AEC's health-studies manager, to release some favorable findings.

The AEC had become alarmed by research being done by Samuel Milham, researcher in the state of Washington's Department of Health. In looking at the death certificates of some 300,000 workers in Washington, Milham noticed what appeared to be an inordinate number of cancer deaths among former workers at the Hanford facility, the first nuclear-weapons center in the country. In his concern, Milham notified the AEC, and following their wishes, refrained from publishing the information he appeared to have uncovered. When the AEC asked Mancuso to discredit Milham's work, Mancuso refused. The government then turned to Batelle Pacific Northwest Laboratories for the "right" answer; it was forthcoming from Batelle (whose contracts with the government were numerous). Batelle's judgment was that Milham's work reflected statistical bias.

Meanwhile, Mancuso continued his work until he was summoned in March 1975 and told that his contract would be canceled as of July 1977. Although his most recent peer review had been excellent, Mancuso's work had become unimportant. The transfer of the study to the federal government's Oak Ridge, Tennessee, staff prompted a letter of protest to James Schlesinger, then secretary of the Department of Energy, on the part of the Environmental Policy

Center; Friends of the Earth; Oil, Chemical and Atomic Workers; the Environmental Defense Fund Public Interest Research Group; the Sierra Club; the Union of Concerned Scientists; and the Natural Resources Defense Council. It was to no avail. Rather than bowing out gracefully, however, Mancuso carried on his studies with the additional help of an internationally respected epidemiologist, Alice Stewart, and her associate, a biostatistician, George Kneale. The closer they looked, the more serious the threat of low-level radiation appeared.

The rate of cancer deaths of workers at the Hanford facility was 7 percent higher than expected, although the average worker only received annual doses of two rem a year, well below the government's five-rem acceptable limit for workers. If that phenomenon were to hold true for workers at other nuclear facilities, government standards might have had to become tougher, not a welcome prospect for the nuclear power industry. Mancuso wanted to make his findings public, but the government agency, then known as the Energy Research and Development Agency (ERDA), wanted him to hold off. Mancuso understandably maintained that the best way to determine whether the Hanford findings were aberrations would be to study other facilities and their workers in equal depth. He wanted his contract renewed to do just that. The ERDA was not interested. His contract was not renewed, first on the undocumented grounds that there were deficiencies in Mancuso's performance, then on the grounds of his supposed imminent retirement, though he was still eight years away from the age of mandatory retirement.

Mancuso's findings were indeed controversial. The federal government felt that they were grounded on insufficient data; but John Gofman, after ten months of studying the data, largely concurred with Mancuso's conclusions. What is disturbing is the way Mancuso's inquiry was abruptly terminated after a decade of painstaking work. The government's actions throw into question its openness to pursue an avenue of research that might potentially be harmful to the nuclear power industry. Particularly in view of what is at stake—namely, the safety and health of workers in nuclear power plants—the government's refusal to allow Mancuso to carry on his work appears to be irresponsible.

In addition to understanding the resources and motivation of the nuclear power industry in selling this technology, it is important to look at the consumer of electricity and attempt to understand the basis upon which a decision on this issue is made, as well as how that decision can be expressed.

BUYING THE IDEA OF NUCLEAR POWER

Because those who provide electricity enjoy the benefits of a monopoly in its delivery, it is virtually impossible for consumers to express a preference in the marketplace as to how electricity should be generated. Realistically, people opposed to nuclear power cannot boycott electricity. Thus, any expression of preference must be made in a political way, as occurred in Seattle, for example. The citizen as a political force supplants the consumer as an economic force in this issue. Even if a clear majority opposed nuclear power, those with that view would have to work through the political system rather than make a simple decision not to buy the product.

There are several reasons why people have accepted nuclear power as readily as they have.

1. Deference to technical expertise: Nuclear power is an intimidating subject, and many people do not feel qualified to make a judgment on it. This fact, coupled with the one-sided nature of information received by the public through advertisements on television and radio, in newspapers and magazines, combine to make this technology as acceptable as it is. There is still a fair measure of confidence in the integrity of corporate bodies of the kind that generate and sell electricity. Yet that confidence is on occasion misplaced; this issue has been known to be misrepresented by utility companies. This leads to the second main reason nuclear power is accepted by much of the public.

2. Information gaps in the public: Because the public still has much to learn about this subject, it is largely at the mercy of the press and the power companies for the information it gets. This provides the latter an opportunity to present half-truths, which are factual but highly misleading and which distort one facet or another of the issue. An example of this already referred to is the federal government's definition of <u>high-level waste</u>, which does not even include spent fuel rods.

3. The mentality of energy scarcity: There is a strong and understandable sentiment that no potential energy source should be foreclosed in the United States at this time. In the case of generating and using electricity, however, there are many other options that exist, but which have not been used to a fraction of their potential. Because a technology such as cogeneration accounts for a minor amount of electrical power now, its full potential is not appreciated. Nuclear power plants, on the other hand, are visible; phasing them out, therefore, represents a threat to an existing en-

ergy source. Even though conservation and alternative sources of power could easily and economically fill the void, there is naturally some reluctance to move toward a new way of doing things. Some utility companies nurture the idea that we can either have nuclear power and enough energy or face the prospect of brownouts and blackouts. This leads to the next point.

4. False choices: The advertisement by Middle South Utilities shows a clear example of false choices being given to citizens. The implication of the advertisement is that coal and nuclear power are the only primary, immediately available ways electricity needs can be met. The advertisement ignores conservation and cogeneration and the multitude of other alternatives for meeting electrical power needs. The importance of the presentation of false choices by supporters of nuclear power cannot be overstated. Nuclear power advocates often suggest that the choice is between nuclear power and inadequate amounts of electricity; between clean and cheap nuclear power and coal; between nuclear power and even greater dependence on the Middle East for oil; between nuclear power and economic stagnation. To the extent that the choices are limited in that manner, the desirability of nuclear power is deceptively exaggerated.

5. Faith in government: Despite all the skepticism and negative attitudes that many people have come to express toward government, there is a sense that regulatory authorities and government leaders would not allow unnecessary risks to be run with the public health and welfare. There is an assumption that if there were safer, cleaner, and cheaper ways of meeting the nation's electrical needs, the government would be forcefully pushing those ways. What the general public overlooks is that in this case the government itself developed the technology in question and thus cannot be completely objective in its assessment of the risks and benefits.

All of these factors make the idea of nuclear power still acceptable to the U.S. public. To be sure, the accident at Three Mile Island has done much to alter public opinion and has revealed the shallowness of popular support for this way of generating electricity. For the reasons already listed, however, there is no consensus on this issue. Some observers had hoped that the President's Commission on the Accident at Three Mile Island would reach a more broadranging verdict on nuclear power in this country. It chose only to deal with the more narrow issue of the deficiencies in the regulation and functioning of one nuclear power plant as revealed by the accident. The commission really did little to answer the question of what the United States should do in the future about nuclear power. That issue should be addressed in a systematic way that assesses both the nation's resources and needs for electricity.

ENDING HIDDEN POWER

The decision on the future of nuclear power is too important to be left to the nuclear power industry alone. Because it has a considerable impact on the public, this issue must be addressed in such a way as to involve as much of the public as possible, as openly as possible, with as much access to good and accurate information as possible. Most important, the real choices must be presented to the public and their policy makers, not the false choices that so often characterize the current discussion of nuclear power.

The first step should be to define the question. Here the example of Seattle should be followed by the nation: we must assess our true needs for electricity, as well as the resources and the most appropriate technologies to meet those needs, based on as complete a presentation of the facts as is possible. Even supporters of nuclear power should be willing to take their chances with that kind of mission. If the focus of the question is not merely nuclear power but, rather, the total electrical needs and resources, the true alternatives are likely to surface in a constructive and enlightening way. Two important considerations must be kept in mind when we assess how this decision should be made: our choices involve both rational decision making and the expression of values of the U.S. people. Whatever public mechanism is created to deal with this issue must accommodate both. I would recommend that the job be done through two bodies—a science court and the Congress of the United States.

The idea of a science court is not new. For example, Arthur Kantrowitz, head of Avco Everett Research Laboratory in Everett, Massachusetts, and a key engineer in the U.S. space program, has been advocating it for a long time. The science court would actually conduct a trial to determine the facts of this technical and controversial subject. The jury would be made up of scientists who are not members of the nuclear power industry and not identified with the antinuclear movement. Those holding both points of view would be able to submit testimony, which the technically expert jurors could evaluate. A judge would preside and ensure that the discussion and debate remained relevant and fair. The science court would be empowered to subpoena records from the nuclear power industry, including the federal government, as well as any other source deemed necessary. The impact of rhetoric would thus be minimized, and an honest effort to ascertain the relevant facts could be made. The science court would present facts on the electrical needs and resources in the United States as well as estimated costs and risks associated with each option, including nuclear power. It could recommend an appropriate strategy for the future, but it could not make a decision. These findings of fact and recommendations would be transmitted to the Congress and to the president.

Ultimately, the decision on nuclear power should be made by elected officials who reflect the values of the people. State legislators and governors should be able to decide whether, and under what conditions, nuclear power should be allowed within their boundaries. But it is also appropriate for the national political leadership to review the findings of fact provided by the science court, and based on the evidence and the alternatives decide whether nuclear power should be allowed to grow. Currently, nuclear power accounts for 13 percent of the nation's electrical energy, but the nuclear power industry hopes to double that amount in less than a decade. The Congress and the president must address the issue forthrightly, openly, and quickly. That could be the beginning of an effective plan and strategy for attaining a nonnuclear future in the United States.

There is another kind of hidden power of critical importance: the potential power of the individual to help reverse the trend toward costly giantism and centralization in the generation of electricity. If the government seems at times as remote and inaccessible as a massive generating plant, the power of its leaders still comes directly from citizens—individual citizens whose views and opinions are critically important to the shaping of policy. Moreover, potential power of concerned citizens goes well beyond the impact of writing one's congressman, important though that is. That power involves learning what alternatives are available now at what cost and with what possible financing. The U.S. Department of Energy and its counterparts in the various states can help provide that information. That power involves learning about the tax advantages of installing conservation devices or alternative energy systems. The Internal Revenue Service and one's state department of revenue can help there. That power involves community organization for conservation or the installation of appropriate decentralized alternative technologies where possible. In the end, the final power to decide the future of nuclear power in the United States resides where it must in a democracy: that power rests in the hands of the people.

NOTES

1. Dan Dorfman, "For Exxon's Eyes Only," Esquire, June 19, 1979, p. 16.
2. Ralph Nader and John Abbotts, The Menace of Atomic Energy (New York: W. W. Norton, 1977), pp. 113-19.
3. Howard Kohn, "The Government's Quiet War on Scientists Who Know Too Much," Rolling Stone, March 23, 1978, pp. 42-44.

REFERENCES

Articles

Dorfman, Dan. "For Exxon's Eyes Only." *Esquire*, June 19, 1979, p. 16.

Kohn, Howard. "The Government's Quiet War on Scientists Who Know Too Much." *Rolling Stone*, March 23, 1978.

"Weighing the Evidence." *Time*, February 23, 1976, p. 45.

Book

Nader, Ralph, and John Abbotts. *The Menace of Atomic Energy*. New York: W. W. Norton, 1977.

APPENDIX A

A LOOK AT ATOMS AND RADIOACTIVE WASTE

In order to understand the nature of radioactive waste, it is helpful to know about the process that produces it. All matter is made up of atoms, each of which consists of a positively charged nucleus made up of protons and neutrons surrounded by a cloud of negatively charged electrons. The number of protons in the nucleus determines which chemical element an atom is—whether hydrogen, carbon, or another of over 90 such elements. Uranium, the heaviest of the natural chemical elements, has 92 protons in its nucleus. Weight variations between atoms of the same chemical are called isotopes, and they come about because of differences in the number of neutrons in the nucleus. For example, uranium 238 has 238 protons and neutrons, while uranium 235 has three fewer neutrons.

Nuclear fission occurs when the nucleus of an atom is split into two or more smaller atoms. An atom with an unstable nucleus like that of uranium produces vast amounts of energy when it is split, energy that had been used to bind the nucleus together. The energy is emitted in the form of electromagnetic and particulate radiation. (The resulting smaller atoms will always be elements other than uranium, since the nature of the element-determining nucleus will have to have changed as a result of the fission.) Not all matter is fissionable, but uranium in some forms is; because it is used in commercial nuclear power, uranium deserves a closer look.

When the nuclei of certain kinds of atoms, such as those of uranium, spontaneously disintegrate, they are referred to as radioactive isotopes or radioisotopes. That disintegration results in the emission of particles and rays termed, collectively, radiation. There are hundreds of radioisotopes; some, like U-235 and U-238, occur in nature, and others, such as plutonium 239, are man-made. Materials containing such radioisotopes are said to be radioactive. The radioisotope U-235 is the most commonly used fissionable material in nuclear power, and less than 1 percent of uranium in nature is of this special kind. Fission of a nucleus of U-235 occurs when it absorbs an additional neutron and splits into two fragments, which may themselves be unstable and undergo decay, releasing additional energy. The nuclear chain reaction that generates great energy and heat comes about as a result of the neutrons that escape the fissioned atom, in turn splitting other atoms.

Not all uranium is fissionable, and even some U-235 fails to fission during a nuclear reaction when it simply absorbs the entering neutron. This has the effect of increasing the atomic mass by one, changing the material from U-235 to U-236. In the case of U-238, an absorbed neutron will create plutonium 239, and the by-product is called transuranic waste, because it has greater atomic mass than uranium. On the other hand, the fragments that are formed by the splitting of the atom are called fission products, another form of nuclear waste.

The length of time radioactive waste remains dangerous centers on the decay rate each radioisotope has. That decay rate is called the half-life of the radioisotope and is defined as the time required for one-half of the nuclei in a sample of material to undergo disintegration. A long half-life means a potentially long period of hazard for living things. The following is a list of half-lives of some radioisotopes:

	Years
Americium 241	458
Americium 242	152
Carbon 14	5,770
Iodine 129	17,200,000
Iodine 131	0.02
Phosphorus 32	0.04
Plutonium 239	24,360
Plutonium 241	13
Radium 226	1,600
Strontium 90	28
Tritium	13

Source: Handbook of Chemistry and Physics (Chemical Rubber, 1968).

capital. The net-income figure finally arrived at will be used later to calculate revenue deficiency.

The average rate base is like the asset side of a balance sheet. It is comprised mainly of the value of the utility plant in service, computed by averaging the plant costs of the utility. From that sum is deducted the accumulated depreciation, to reach a figure for net utility plant in service. Add to that the value of construction work in progress (plants not yet working but being built) and working capital (set in Minnesota for large utilities on the basis of "lead-lag studies," which assess when the utilities get paid by consumers and pay their creditors in turn). Subtract from that figure the accumulated deferred income taxes, sometimes called phantom taxes because they are bookkeeping numbers, not taxes paid out. The grand total of all of those computations is the average rate base. It will soon become evident that the larger the rate base, the greater the income allowed to the utility. There is thus little incentive, at least from the standpoint of generating income through the rate-making process, for the utilities to keep the cost of new power plants as low as possible. On the contrary, the larger the rate base, the larger the amount of required operating income.

Another element that must be calculated to come up with the revenue deficiency is the allowed rate of return on capital sought and given, the most important issue to be decided by the commission. The capital structure of a utility is made up of common equity, long-term debt, short-term debt, and preferred stock. Since the last three elements are already set, the key focus of debate and discussion is on the return on common equity. A crucial consideration for the public service commission is the effect a given rate of return will have on the utility's ability to attract more investment capital or obtain funds from loans and bond issuance. Hence, a utility would maintain that if the commission were to allow a high return on equity, other capital presumably can be made available to the utility in a way that benefits the consumer more in the long run. Utilities generally seek a return on common equity in the range of 12 percent to 15 percent. Subjective and objective criteria are applied to the decision-making process, and the allowed rate of return is viewed in the context of what other, similarly sized utilities are getting. Clearly, public decisions in this part of utility regulation have an effect on what private investors and lenders will do, and the public is therefore a strong partner with the utility in assuring the utility's economic health.

When the rate of return is established, a calculation that involves estimating the weighted cost of capital for short-term and

APPENDIX B
UTILITY REGULATION AND RATE MAKING

Utilities are regulated monopolies, and the
uct—in this case, electricity—is set not by the fo1
petitive marketplace but by what the utilities are
regulators in the various states—often called the
commission or public utility commission. A bri(
are set in Minnesota might be helpful in understa
generally, since there are rarely major discrep&
states in how this is done. To be sure, a state l
has a publicly owned system of electrical genera
regulation somewhat differently from other state
ity is delivered by utility companies either. Ru1
Administration cooperatives and municipal syste
service. In the case of generation and transmis
there is usually not any state government rate r

The complicated process of rate making c
determine the revenue deficiency a utility comp
caused it to come to the public service commis:
increase. In reaching that and other intervenin
commission looks at three different financial s(
ing income statement, the average rate base, a
determines the revenue deficiency.

The operating income statement determir
utility. Operating revenues are computed on tl
rates, multiplied by sales according to custom
home, and the like). A critical fact is that ge1
low utilities an automatic adjustment clause, v
documented, permits an automatic passing on
cost increases for fuel. Thus, if the cost of u
creasing at a particularly fast rate, this provi
tion will take its toll directly and automaticall

Expenses are subtracted from operating
with net income. Expenses include the cost o
tenance expenses, the cost of fuel, administra
insurance and advertising, and almost all oth
terest expenses on the debt, which is figured

long-term debt, as well as common and preferred stock, that rate of return is multiplied by the average rate base to come up with the required operating income. From that is subtracted the net income (which has already been defined) to come up with the income deficiency. Then a figure is calculated to factor in federal, state, and local taxes; this is called the gross revenue conversion factor, and it is multiplied by the income deficiency to determine the revenue deficiency, the amount that needs to be added to current rates to establish the new rates.

The Public Service Commission then determines how much of the higher rates each class of customer must pay.

APPENDIX C
IONIZING RADIATION
AND ITS EFFECTS ON CELLS

The danger of exposure to radioactive substances centers on their effects on human cells, the important building blocks of life. One gram of tissue contains about one billion cells, and the wrong kind of injury to just one of them can eventually lead to cancer. Radiation is one of many ways the human-cell structure can be distorted, damaged, or destroyed; for our purposes it is important to look at what is known about the effects of exposure to radioactive materials.

Radioactive materials emit ionizing radiation, so called because this kind of energy separates electrons from atoms, producing ions, or atoms with an electrical charge. There are two forms of ionizing radiation: electromagnetic radiation, similar to visible light (but with a much greater level of energy), which includes X rays and gamma rays, and particulate radiation, which includes alpha and beta particles. Ionizing radiation of any kind can damage cells, the amount of the damage depending largely on the dose of radiation received.

Dosage of radioactivity is measured in rads and rems, terms describing roughly equivalent exposure to radioactive materials. A rem (roentgen equivalent in man) is a measure of the number of electrons torn away from molecules by a beam of radiation, and a rad (radiation absorbed dose) measures the amount of absorbed radioactive energy. In discussions of allowable exposure, the units used most often are millirems or millirads, measuring one-thousandth of a rem or rad.

Background radiation from cosmic rays and naturally occurring radioactive material on earth accounts for the single largest source of exposure. In 1979 the Interagency Task Force on the Health Effects of Ionizing Radiation estimated that background radiation accounts for 20 million person-rems annually in the United States, or roughly 100 millirems per person. Advocates of nuclear power often cite this fact as a means of trying to demonstrate how inconsequential commercial nuclear power is in adding to the total amount of radioactive exposure people experience. A convincing case can be made that radiation from X rays has done considerably more damage to human cells than radiation coming out of a safely functioning nuclear power plant. However, the fact that radiation comes from many sources and can be dangerous in many different forms does not diminish the importance of understanding the danger of low-level radiation, and clearly nuclear power plants contribute to such low-level radiation.

Depending on the circumstances, radioactive material can be inhaled, swallowed, or taken in through the skin. It can damage internal organs or the skin. The nature of a person's exposure is of critical importance. For example, alpha particles, which are electrically charged nuclei of helium atoms, are quite large and therefore do not have the penetrating ability of beta particles or gamma rays. Alpha particles transfer so much of their energy in a short distance that a person's skin is sufficient to protect him from radioactive damage to internal organs. But if alpha-emitting radioactive material is inhaled, it can be distributed along the respiratory tract and irradiate those cells that are especially prone to develop cancer. It is just such inhalation that helps explain the high incidence of lung cancer among uranium miners.

In addition to how radiation is absorbed, the kind of radiation absorbed is also important. While no kind of radiation is safe, be it in the form of X rays, gamma rays, or alpha or beta particles, strong evidence presented by Donald Geesaman of the University of Minnesota and Arthur Tamplin indicates that "hot" particles made up of alpha-emitting substances can be from 10 to 1,000 times more effective in causing cancer than would be the case if the same number of rads were delivered more diffusely to an organ such as a lung. Plutonium 239 is a notable example. When burned, fine particles of pure plutonium 239 oxide are intense sources of alpha particles. Hence, P-239, once heralded as the fuel of the future when reprocessing was allowed, is one of the most potentially lethal substances on earth. With a half-life of 24,000 years, plutonium oxide of this kind could, in the event of a calamity, be spread around the earth, resuspended in the air, and remain deadly for 100,000 to 200,000 years. This kind of consideration justifiably weighed heavily in the decision to discontinue reprocessing and thereby reduce the amount of P-239 that is circulating in our country.

Ionizing radiation causes a disorganization of biological cells, which are otherwise remarkably well-organized accumulations of chemical substances. If the damage is adequately catastrophic, the cells simply die. If they are merely injured, they can continue to live, divide, and reproduce new cells, possibly new cells that are similarly injured. Fatal injury to cells can actually be less dangerous to the body than nonfatal injury to cells because of the body's natural ability to reproduce normal cells. A nonfatal injury to just one cell has the potential of starting cancer or leukemia. Furthermore, because of the long latency period involved, there is no way to identify exactly when the process began. For that reason, it can be said that the long-term effect of radiation is an insidious and real hidden power not fully comprehended at the time of exposure by the people exposed to it.

Age is critically important in determining the effect of ionizing radiation on human beings. Fetuses and young children appear to be much more vulnerable to radiation exposure than are adults. A report from the Department of Health, Education and Welfare states the following.

> Developmental effects are abnormalities that occur in individuals who were exposed to radiation while in the womb or as young children. Animal experiments suggest that doses as low as 5 rem in early pregnancy may produce increased incidence of skeletal, nervous system and other malformations. Doses from 10 to 19 rem to human fetuses have been shown to produce small head size, and doses above 150 rem have been associated with mental retardation. Doses in the 0.2 to 20 rem range appear to increase the risk of childhood cancer.[1]

It is believed that a given amount of radiation increases the risk of future cancer or leukemia 50 times more for an embryo during gestation than for an adult. This would seem to suggest that women of child-bearing years would be well advised to avoid any occupation that would expose them to unusual levels of radiation. Beyond that, society should weigh carefully the potential risks in the proliferation of any kind of radiation-increasing technology. The future, as well as the present, is at stake.

Excessive radiation can have a direct bearing on future generations. To understand why, it is necessary to look further at the structure and function of cells. Each cell has a nucleus, which is considered a crucial site for cell injury by ionizing radiation. Furthermore, the chromosomes in each cell are critically important in understanding the impact of ionizing radiation. In each normal human cell—except for certain stages of sperm and ova cells—there are 46 chromosomes. The chromosomes carry the information in the cell that determines the production of proteins for cell function, growth, and division. Broken or divided cells can be seen with a microscope. Additional radiation-induced injury—beyond what is visible under a microscope—also occurs. Some form of chromosome injury is widely assumed to cause the cancer associated with high levels of radiation.

There are two primary ways that the cell's information system can be altered by exposure to radiation. Radiation can alter the chemical makeup of a single gene, that element by which hereditary characteristics are transmitted and determined. When this occurs it is called a point mutation. The other way the cell can be damaged is when the radiation breaks off a piece of the chromosome, causing it and its hundreds of genes to be lost in the future division of the cells and the creation of daughter cells. This kind of mutation is called, appropriately, a deletion. It is widely believed that a single radiation event can cause enough chromosomal change to cause cancer under the right circumstances.

As we have noted, there are two cells that do not have 46 chromosomes within them: those that produce sperm, called spermatogonia, and those that produce ova, called oocytes. Mature spermatozoa and mature ova each have 23 chromosomes. Injury to either the sperm or ova chromosomes by either kind of mutation can be carried forward into every cell of a new human being and in turn be transmitted from generation to generation. While some mutations affecting the sperm and ova may be serious enough to make reproduction impossible, many such mutations do permit the development of human beings afflicted with fundamental genetic disadvantages. One of the many unknowns in a discussion of the effects of ionizing radiation on people is how many such mutations can occur in the human population before such mutations become a real threat to the successful propagation of future generations.

Geneticists have voiced concern about the danger of allowing any increase in the rate of mutations in the general population. To be sure, natural radiation can account for some such changes, but the question is raised, Do current radiation standards associated with the generation of nuclear power provide adequate protection against serious changes in the rate of mutation? One distinguished geneticist and Nobel laureate, Joshua Lederberg of Stanford University, emphatically feels that is not so. In an affidavit before the Vermont Public Service Board in 1970, Lederberg said that government radiation standards allowed for a 10 percent increase in the mutation rate, far too high in his judgment.[2] Of course, there are differences of opinion on what an acceptable level should be. It would seem prudent, however, to err on the side of caution in view of how little is known about the sources and causes of genetic mutation throughout our environment. Furthermore, the welfare of people in the distant future is at stake. Perhaps one can make a case for a technology that provides danger only to those who use it; but it is far less easy to defend any technology, such as nuclear power, which potentially has such long-term, disastrous effects.

NOTES

1. U.S., Department of Health, Education and Welfare, <u>Report of the Interagency Task Force on the Health Effects of Ionizing Radiation</u>, Washington, D.C., June 1979, p. 31.

2. Joshua Lederberg, "Affidavit before the Public Service Board of Vermont," Docket no. 3445, September 1970.

INDEX

Abbotts, John, xiv, 149
Abrahamson, Dean A., 6, 9, 109
American Wind Energy Association, 121
Atomic Energy Act, 88
Atomic Industrial Forum, 146, 149
Avco Everett Research Laboratory, 154

Babcock and Wilcox, 45, 81, 82, 86, 90
Badger Safe Energy Alliance, 32
Batelle Pacific Northwest Laboratory, 17, 37, 40, 42, 43, 44, 48, 150
Berger, John, 126, 129, 130
Berlin, Edward, 104
biomass: wood, 124, 136; refuse, 125
Bonneville Power Co., 129
breeder reactor, 31, 80
Browns Ferry nuclear plant, 80, 82
Burlington Electric Co., 124

California Public Utility Commission, 116
Carlson, Richard, 118, 131-35
Carter, Jimmy, 4, 9, 10, 11, 91, 127
Chapman, Duane, 45
Cicchetti, Charles, 104
Claude, George, 130
Climax Uranium Co., 64

Code of Federal Regulations, Title 10, 88
Coffin, Ned (chairman, Enertech Corp.), 121
cogeneration, pp. 118, 120, 133, 135, 136, 152
Commoner, Barry, 127
Connecticut Resource Recovery Authority, 125
costs of nuclear power: accident, 45-47, 48-49, 149; capital, 23, 26-28, 47-48, 117; compared with coal, 24, 147-49, 153; enrichment services and, 41-42; fuel, 23, 30-32; future, 32, 35-36, 48; government subsidies for, 24-25, 36-37, 45, 145; in context of regulated monopoly, 25-26, 48; and liability insurance (Price Anderson Act), 42-43, 45, 48; regulation, 42; research-and-development, 37-40; and uranium-industry subsidies, 40-41
Crystal River nuclear plant, 92

Davis-Besse nuclear plant, 81
Dawson, James, 84
DeBuchannane, George, 13
decommissioning nuclear plants, 16-17
demand for electricity, estimated, 131-32
Dorfman, Dan, 147
Dow Chemical Co., 118

Dowd, James and John, 123
Dresden nuclear plant, 94

Edison Electric Institute, 147
Eisenhower, Dwight D., 9
electrical energy conservation, 103-18; as affected by pricing of electricity, 103-6; as affecting utility companies, 112-16; Los Angeles example of, 106-7, 111; organization of, 111-12; Seattle example of, 107-8, 111, 136, 154; technologies of, 108-10
electrical power rate regulation, 161-63
Energy Tax of 1978, 120
Enrico Fermi nuclear plant, 80, 82
Environmental Defense Fund Public Interest Research Group, 151
Environmental Policy Center, 150-51
excess U.S. generating capacity, 117, 130
Export-Import Bank, 43
Exxon Corp., 110, 146, 147, 149

Federal Radiation Council, 57
Freedman, David, 118, 131-35
Freedom of Information Act, 150
Freeman, S. David, 104, 115
Friends of the Earth, 151

Geesaman, Donald P., 6, 9, 166
General Electric Corp., 44, 145, 146
General Public Utilities (GPU), 45
geothermal power, 125-26, 136
Getty Nuclear Fuel Services, 67
Gillen, William, 104
Gofman, John, xiv, 57-58, 69, 151

Harrisburg, Pa., 45
Hartley, Fred L., 125-26
Hellman, Richard, 147
Heronemus, William, 129
Hutton, E. F., Co., 112
hydroelectric power, 107, 122-24, 133, 135, 136

Institute for Electrical and Electronic Engineers, 57

Johns Hopkins University, 61

Kantrowitz, Arthur, 154
Kneale, George, 58, 151
Krugman, Hartmut, 4, 6
Kuhns, William, 45

Lawrence Radiation Laboratory, 57
Lederberg, Joshua, 168
Leuthold, Steve, 46
Lewis, H. W., 94
Lewis Report, 94-95
Lovins, Amory, 103-4, 110, 112, 113-15
low-level radiation, 55-69; alpha particles and, 56, 166; beta particles and, 166; dangers of, 55-69, 150-51, 165-69; gamma rays and, 59, 60, 67; genetic effects of, 59; measurement of, 56-57, 62; safety precautions and, 56; standards of, 57-59, 59-60, 62-63; X rays and, 56, 59, 60, 165
Luce, Charles (chairman, Consolidated Edison of New York), 116

Mancuso, Thomas, 58-59, 61, 150-51
Marine Midland Bank of New York, 123

INDEX / 173

Marks, Sidney, 150
Massachusetts Institute of Technology, 93
McPhee, John, 123
meltdown of nuclear reactor, 77, 80, 81, 92
Metropolitan Edison Co., 82
Meyer, Eugene (vice-president, Kidder Peabody Co.), 116
MHB Technical Associates, 15
Michelson, Carlyle, 82
Middle South Utilities, 148, 153
Mielke, Frederick (chairman, Pacific Gas and Electric Co.), 116
Milham, Samuel, 150
Miller, Saunders, 31-32
Minnesota House of Representatives, xi, xiii, 49, 97, 136; Energy and Utilities Committee of, 1, 55
Minnesota Public Interest Research Group (MPIRG), xii
Monticello nuclear plant, xii, xiii, 3, 55
Moody's Investors Service, 46

Nader, Ralph, xiv, 149
Nadis, Steve, 130
National Academy of Sciences (NAS), 59, 60
National Council of Radiation Protection and Management, 63-64
National Electric Reliability Council (NERC), 131, 132, 133
National Energy Act of 1978, 123
National Uranium Resource Evaluation (NURE), 43
Natural Resources Defense Council, 151
New England River Basins Commission, 122
New Mexico, 8-9, 65-66

Northern States Power Co. (NSP), xii, xiii, 1, 2, 18, 24, 32, 33, 84, 124, 129
nuclear fission, 11, 45, 159-60
nuclear fuel production, 23, 31, 32, 47-48, 56, 61-62, 67
Nuclear Fuel Services (NFS), 15
nuclear fusion, 101
nuclear power plants: accident prevention in, 74, 83, 91, 96; malfunctions at, 74, 77-84, 96; reactor operations in, 74-77; regulation of, 74, 87-91, 96; risk assessment of, 93, 97, 149; safety monitoring of, 74, 83-84, 87-91, 96; training workers of, 74, 84, 86, 96; and uncertainty involved in central-station generation, 117
nuclear waste: constitutional issues of, 8; cost of, 15-16, 18; and disposal technologies and sites, 12-14, 18; high-level, 2-6; military versus commercial volume of, 5; political issues of, 7-9, 11, 12, 13; and toxicity, 6-7; transportation of, 14; volume of, 6-7, 18; and the Waste Isolation Pilot Plant, 8

Oak Ridge Gaseous Diffusion Plant, 67
ocean thermal energy conversion, 130
Oil, Chemical and Atomic Workers union, 151

Pacific Gas and Electric Co., 116, 125
Pacific Power and Light Co., 116
Perception for the Professional, 46
photovoltaic electricity, 125-28, 135, 136

Piper, Jaffrey, and Hopwood, 46
President's Commission on the Accident at Three Mile Island, 73, 83, 85, 87, 91, 153
Public Utility Regulatory Policy Act (PURPA) of 1978, 105, 119-20, 122
Puget Sound Power and Light, 116

radiation and its effect on cells, 165-69 (see also low-level radiation)
radon gas, 62, 63, 64, 65
Rasmussen, Norman, 93
Rasmussen Report, 93-95
risk assessment of nuclear-power-plant accidents, 92-97; concept of necessary risk, 93
Rockwell, Winthrop, 73, 83
Rosen, Benjamin, 126, 127
Ross, Marc H., 108
Rural Electrification Administration, 120

San Onofre power plant, 17
Schlesinger, James, 149, 150
science court, 154-55
Scott, Robert, 118, 131-35
Sierra Club, xii, 32, 35, 94, 147, 151
solar power, centralized, 128 (see also photovoltaic electricity)
Southern California Edison, 129
State Planning Council on Radioactive Waste Management, 10
Stewart, Alice, 58, 151

Tamplin, Arthur, xiv, 57-58, 166
Tennessee Valley Authority, 82, 104, 116
Texas Instruments Co., 126
Thermo Electron Corp., 118
Thompson, Theos, 58

Three Mile Island, xi, 35, 45-47, 48-49, 56, 73, 74, 81, 82, 83-84, 85, 89, 90, 91, 92, 95, 102, 146, 153
Tyrone power plant, 32-36; capacity-factor assumptions, 34; discount-rate analysis, 33-34, 36; escalation-rate assumptions, 33-34; rejection, reasons for, 35

Union Carbide Corp., 146
Union of Concerned Scientists (UCS), 94, 147, 151
United Nuclear Corp., 64, 65
U.S. government, agencies and departments of: Agency for International Development, 43; Army Corps of Engineers, 122, 133; Atomic Energy Commission (AEC), 9, 41, 42, 57, 58, 66, 89, 145, 149, 150; Central Intelligence Agency, 149; Department of Energy (DOE), 9, 12, 15, 30-31, 36, 43, 58, 110, 118, 127, 128, 146, 149, 150; Department of the Interior, 10, 13 [Geological Survey, 10, 13]; Department of Transportation, 10; Energy Research and Development Agency (ERDA), 9, 151; Environmental Protection Agency (EPA), 10, 66; Federal Energy Administration (FEA), 127; Federal Energy Regulatory Commission (FERC), 119, 123; Federal Power Commission, 130, 147; General Accounting Office (GAO), 42, 86, 87, 88-90; Interagency Task Force on the Health Effects of Ionizing Radiation, 60-61, 65, 67-68, 165; Internal Revenue Service, 155; Mine Safety and Health Administra-

tion, 62, 63; National Aeronautics and Space Administration, 93, 110; Nuclear Regulatory Commission (NRC), 4, 9 15, 17, 42, 65, 67, 73, 77, 79, 81, 85, 86-91, 94, 96, 146
University of Wisconsin, 109
Upton, Arthur, 58
uranium: enrichment, 31, 66-67; fuel fabrication, 31, 67; milling, 31, 64-66; mining, 31, 62-64, 69; transportation, 67-68

Vermont Public Service Board, 168
von Hippel, Frank, 4, 6

WASH-740 (nuclear safety study), 149
Washington Public Power Supply System, 107
Western Wisconsin Utilities (WWU), 32, 35
Westinghouse Corp., 63, 146; Nuclear Training Center, 84-85
West Valley, N.Y., 15-16
Williams, Robert H., 108
wind generation, 101, 102, 120-22, 136; centralized, 129
Windfall Profits Tax of 1980, 120, 123
Wisconsin Public Service Commission, 32, 35, 36
Wolff, Ben, 121

ABOUT THE AUTHOR

TODD OTIS is a second-term Democratic Farm-Labor party member of the Minnesota House of Representatives. A native of Minnesota, Otis is a graduate of Harvard University and has a master's degree in journalism from Columbia University. His work experience includes two years in the Peace Corps in West Africa, several years in corporate community relations for Minneapolis-based companies, and two years as a lobbyist for an organization of small businesses. He is married and has two children.